PIS FOR PESTICIDES

Dr Tim Lang & Dr Charlie Clutterbuck

In association with The Pesticides Trust

EBURY PRESS
London

First published 1991 by Ebury Press
an imprint of the Random Century Group
Random Century House
20 Vauxhall Bridge Road
London SW1V 2SA

Editor: Esther Jagger
Designer: Bob Vickers

British Library Cataloguing in Publication Data
Lang, Tim
P is for pesticides.
1. Man. Toxic effects of pesticides
I. Title II. Clutterbuck, Charlie
615.902

ISBN 0–85223–968–8

Typeset in Plantin by Textype Typesetters, Cambridge
Printed and bound in Great Britain by
Mackays of Chatham Plc, Kent

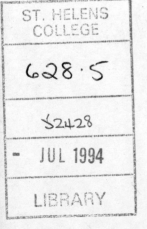

Contents

Acknowledgments

Many people have helped in the making of this book, and we are grateful to all those individuals and organisations, journalists and scientists, who have joined together to share information with the public. We would like to thank: everyone at the Pesticides Trust – Peter Beaumont and David Buffin compiled Tables 1, 3 and 4 and offered invaluable support and advice, though the responsibility for the book is, of course, ours – also, Barbara Dinham and Dave Bull; everyone at Parents for Safe Food, Pamela Stephenson, Olivia Harrison, Gae Exton, Dominique Coles, Louise Baxter, Julia Hunter and all at John Reid Enterprises, Steve Brown, Simon Prytherch, Nikki Neave, and all supporters of Parents for Safe Food; the Principal of Blackburn College, and all friends and colleagues there; the Pesticides Action Network, especially the Europe Office and Saro Rengam of Asia Pacific Regional Coordinating Centre; Peter Hurst of the T&GWU (Transport and General Workers' Union); Michael Cunningham, ex-NUPE (National Union of Public Employees) and Bronwen Bernard of NUPE; Dave Thomas of the Agricultural Trade Group of the T&GWU; Nigel Bryson of BFAWU and many other pioneering trade union members and officers, such as Reg Green and Barry Leathwood; David Gee and Andrew Lees at Friends of the Earth; Nigel Dudley, all at ERR (Earth Resources Research) and the Soil Association; Dr Alan Gear and Jackie Gear of the HDRA, and all who research practical alternatives to pesticides; the Women's Environmental Network; the London Food Commission (1984–90: RIP), particularly Peter Snell; the Food Commission UK Ltd; London Hazards Centre; *The Food Magazine*; the chemicals working group of the European Ecological Consumers' Coordination, and particularly Die Verbraucher Initiative in Bonn; also the Asia and Pacific Region of the International Organisation of Consumers' Unions; Louise Sylvan, Yong Sook Kwok and Roger Westcombe at the Australian Consumers' Association; Craig Merrilees and the National Toxics Campaign in the USA; also all at the US Natural Resources Defence Council; a fine group of independent scientists: Dr Alistair Hay, Dr Andy Watterson, Dr Erik Millstone and Dr Melanie Miller for much encouragement over the years; also to Liz Castledine and Frances Pickles; all at Ebury, but especially Gail Rebuck, who first inspired us, Fiona MacIntyre, our editor, Joanna Sheehan, Deborah Cattermull and Esther Jagger for their patience and help.

This book was written in Nelson, Lancashire, at four addresses in London, partly in Australia and partly in the West Indies. We would like to thank all those anonymous people who produced our computers, printers and disks, and the Post Office whose service made the transfer of data on disks so quick and efficient.

Finally we want to thank all those people who have helped us but who must remain anonymous. They work at the sharp end: in the UK Ministry of Agriculture, Fisheries and Food (MAFF), in the UK agrochemical and farming industries, as City analysts and journalists. Their encouragement and support are testimony to the recognition inside and outside the agrochemical treadmill that change is necessary, overdue and inevitable. Public health, consumer and environment protection are concerns which can override immediate vested interest. All power to your elbows.

Tim Lang
Charlie Clutterbuck

London and Lancashire
January 1991

How to Use This Book

P*art 1* (Chapters 1–17) answers all the main questions we get asked about pesticides, like 'What are they?', 'What foods will I eat them on?', 'Does the label on a product tell me anything?', 'Will they give me cancer?', 'What can I do?', 'Who can help me?' and 'What legal rights do I have?'.

Part 2 gives you at-a-glance information. Let's say you have a bottle of pesticide. You've learnt how to read the label (see Chapters 2 and 3). Now you want to look up more information on this product's chemical ingredients. Here's how to do it.

- Table 1 gives you some key foods and food groups you will find pesticides in, and tells you what you can do about them
- Table 2 lists most commonly used pesticides and gives you detailed information on what they are, together with our advice about them
- Table 3 lists all the hundreds of pesticides approved for agricultural use by the British government. We also tell you how other countries and organisations disagree with our Ministry of Agriculture, Fisheries and Food's diagnosis.
- Table 4 lists all the symptoms from poisoning of the main groups of pesticides

At the end of the book there are:

- the UK authorities' word-for-word explanation on how dangers are assessed
- A list of useful addresses, including organisations you can turn to for more help
- a list of books that we recommend for further reading
- the European Consumers' Pesticide Charter, which lists sensible reforms we'd like to see
- a glossary of abbreviations and technical terms
- our sources are included here, though the general reader will probably not want to bother with them
- and finally there is a comprehensive index so you can quickly find whatever you want to read about.

Introduction

Pesticides are everywhere. There are over four hundred pesticide ingredients approved for use in the UK, and there are over three thousand different brands on the market. They stare out at you from supermarket shelves, garden centres and advertisements. Flick through the farming press adverts in a newsagent's and you cannot miss them. They are called by weird, stark and aggressive names – Bugoff, Kilfactor 9, Kombat, Killpest. . . . These trade names sell themselves to you as getting rid of nasty little insects, mosquitoes in the night, things that blemish pretty plants.

They promise a lot: 'Use Bugoff and your food will survive assault by bugs.' The message is that without Bugoff you are taking a big risk. At worst, some ads hint, without Bugoff humans will starve. They rarely tell you the dangers and risks from pesticides, let alone the alternatives.

Pesticides are also invisible. You cannot buy any X-ray specs from the chemist which reveal pesticide residues to you. Residues can only be found in a laboratory, with expensive equipment. But you can develop a nose for where pesticides might be, once you know how and where to look, wherever you are. Start with your home. Pesticides are:

- sprayed on flowers you buy to look nice in your room
- in the garden shed, promising to deal with pests on your plants
- in the fridge, cupboard or larder, on or in your food
- in your water, from residues trickling through the soil after farm use
- in the air, as you walk through the countryside watching the farmer spray the crops
- on the pavement and kerb outside your house, which the council sprays to keep weeds down
- in flysprays
- in milk and even breastmilk
- in your blood.

And if you look at work, you will find them there too. You may

- make them
- spray them on pavements
- impregnate materials with them
- keep down fleas and cockroaches that you would not get at home.

And if you go out for a meal, there is often more on the menu than meets the eye – unless you buy food which is guaranteed 'organic' or pesticide-free. And even some of that has been found, in government tests, to contain traces of pesticides. Sprays can drift from the land of a farmer using pesticides on to those of someone who doesn't.

Pesticides are so much part of everyday living that they are taken for granted. A fly in the room – reach for the spray. People like their fruit and vegetables to have no marks – that is one reason, we are told, why farmers spray pesticides.

Yet pesticides are different from other chemicals. They are the only chemicals that, as we have just shown you, are found everywhere. Industrial chemicals are found at work but not elsewhere. Food additives are in your food but not elsewhere. Only a pesticide can be in all places. Perhaps in minute amounts, but nobody has worked out the cumulative effects.

This book tries to help you make them visible. Spotting the pesticides is the first step to controlling or avoiding them. This book has therefore been written with the aim of informing you how to:

- spot pesticides
- control pesticides
- understand and question pesticide science
- make choices about pesticides and avoid them, if necessary
- get alternatives.

Often the first you know about a pesticide is when you read stories in the press about something in apples, water or ointment for your child's hair. It is then called a 'scare story'. While we quote specific cases, we do not set out to scare. Our intention is to tell you the full story about pesticides, and to give you the main arguments about them.

Many people are put off questioning pesticides because they feel faced by a wall of science. All the long names seem so incomprehensible. Science apparently justifies how pesticides were developed, how they get put on the market, how safe they are and even how they get banned. This book tells a rather different story. We show how arbitrary and *ad hoc* pesticide science is. We want you to question that science.

All sorts of people ask us questions. Consumers, trade union reps, environmental groups ask us simple questions like 'Why is so and so banned there but not here?' and 'Why is it they change their minds – one year it's OK, then the next it's "Ooh, sorry about that one!"?' We have given a few examples of that in this book. The more that questions are asked, the more there seem to be.

Did you know that some pesticides approved for use in the UK are defined by other countries and authorities as *known* causes of cancer, *known* to be implicated in genetic mutation, *known* to cause allergies or irritation in animals? We think you have the right to know. Look them up in Table 2 and make up your own mind.

We think it is up to you to decide whether you want a risk. We don't buy the argument that says it is risky crossing a road and therefore a little extra pesticide risk is nothing to worry about. Especially when it comes from somebody in a white coat paid by a chemical company.

We challenge official complacency and secrecy about pesticides. Decisions are constantly being made on your behalf by people you have never heard of. But they never ask you or involve someone who can speak on your behalf in their decision. Yet often they make out that they are 'objective'. This is done to put you off. Who are you to question an expert? Especially when you are not told half the facts. Yet how can a decision be 'objective' if it does not involve all sides? And if a decision is that good, why is it not open to public debate?

That is what we would like this book to lead to: a public debate, not a series of so-called scare stories. We would like you all to join in a debate about pesticide usage everywhere.

PART 1

1
What Are Pesticides?

Pesticides are poisons. They include any chemical used to control a pest or disease.

The different types of pesticide

Pesticides include

- *insecticides* that kill insects
- *herbicides* that kill plants
- *fungicides* that kill fungal diseases
- *rodenticides* that kill rats and mice
- *acaricides* that kill mites
- *molluscicides* that kill snails and slugs
- *nematocides* that kill eelworms and threadworms
- *avicides* that kill birds

The word 'pesticide' also includes substances that

- retard plant growth (*growth regulators*)
- remove leaves (*defoliants*)
- speed plant drying (*desiccants*)
- act as fumigants
- and repel or attract insects (*repellents/attractants*)
- there is a new group called *insect growth regulators* (IGRs), which reduce insect growth in various ways.

Veterinary products may include pesticides, although they are usually considered separately. For instance, sheep dips are classed as veterinary products yet they contain pesticides. Sometimes a pesticide becomes a veterinary product, like when it's used on a trout farm.

There are also products known as *commodity chemicals*. These are sold simply as chemical substances, but can also be used as pesticides. There are some substances covered by the Medicines Act which act like pesticides. Fertilisers and animal feedstuffs, complete with drugs, are not considered pesticides although sometimes their functions overlap. All these substances and pesticides are called *agrochemicals*.

Why are pesticides used?

Pesticides are used to reduce crop loss from disease and pest (plant and insect) attack, both before and after harvest. They are also used for public health to control various human pests and disease carriers. By the very nature of these functions, pesticides are released into the environment often close to people.

In 1989, 29,000 tonnes of pesticide concentrate (active ingredient) was used in the UK. This means that roughly 250 g of neat pesticide was applied for every woman, man and child in the UK. When diluted for spraying, some 4.5 billion litres of pesticide will be sprayed this year (and every year) in the UK. All of that ends up in or on your food, water, air, flowers, pavement. . . .

Are there differences between natural and artificial chemicals?

Many chemicals have been used at various times to control a wide variety of pests. There are three main groups:

Natural substances include nicotine, pyrethrum and derris. But 'natural' does not necessarily mean safe – a lot of natural chemicals like these can be very dangerous.

Inorganic substances are derivatives of sulphur, arsenic, mercury and some other metals.

Synthetic compounds cover the vast bulk of present-day pesticides. There are several major categories of synthetic pesticides, including:

● organochlorines, e.g. DDT
● organophosphates, e.g. Malathion
● phenoxyacetic acids, e.g. 2,4-D
● carbamates, e.g. Aldicarb
● synthetic pyrethroids (derived from pyrethrum).

These compounds are sometimes called *organic*. This is a chemical term to distinguish chemicals made of carbon, hydrogen and oxygen from those of 'mineral' origin, which are known as *inorganics*. But don't confuse the chemical term 'organic' with the modern use of the word, to mean food or farming without pesticides. In mainland Europe they tend to use the term *'biological'* for this kind of farming. We prefer the term *'pesticide-free'*, as that is crystal clear.

More details of these chemicals and how they came to be used as pesticides can be found in Chapter 6.

Names people use

Each pesticide has three sorts of names – *chemical, common* and *trade*. Look at Paraquat (on p. 206). This is the common name and the one probably most familiar to you. Its chemical name is a long, unpronounceable one (1,1'-dimethyl-4,4-bipyridinium ion). The trade name is the one chosen by the manufacturing company (e.g. Gramoxone).

The trouble is that each chemical can be called by different common names, especially in different countries – it can all be most confusing. But attempts are now being made to get everyone to conform. The International Organisation of Standards (known as ISO) was set up to decide, among other things, which common name should be used. So Paraquat is an ISO-approved common name – approved everywhere, that is, except Germany.

But there are other conventions and sometimes people use a 'common' name – like nicotine, which is not adopted by the ISO. It is then called a *trivial name*. The plot thickens.

There are also different world conventions about chemical names, so any chemical substance may have several chemical names. One of the main reference journals that researchers use is called *Chemical Abstracts*. It uses the name 1,1'-dimethyl-4,4-bipyridinium ion for Paraquat. Spot the difference.

Finally, there may be dozens of trade names for the same chemical. Manufacturers want you to buy their brand, not their competitors', so they make up all kinds of chemical combinations with all sorts of weird and wonderful names. Apart from Gramoxone, Paraquat can be found as Dextrone X, Weedol, Dexuron, Scythe, Gramazine, Pathclear, Groundhog, Clean-

Dexuron, Scythe, Gramazine, Pathclear, Groundhog, Clean-
sweep and others. Not content with that, there are also code
names or other names that a pesticide may have if it is used for
another purpose.

The names of pesticides were described in 1974 by the
World Health Organisation as being in 'deplorable confusion'.
Despite some progress in standardisation, we have to say that
many years later it is still deplorable.

The names we use in this book

In this book we will always use the common name – the commonly used common name! It is usual scientific practice to
spell pesticide common names with a small letter, e.g.
paraquat. But to make these names stand out in this book we
have used a capital letter. Because we use these common names
throughout, there will be some pesticides that you will not find
here. A lot of people know the pesticide Gramoxone, for
instance. It is actually the trade name for ICI's brand of
Paraquat. So if you are a gardener and have a container of pesticide and want to read what we think about it, you must look
on the label for the common name, which will always be listed
in the ingredients. On your bottle of Gramoxone you would
find the word 'Paraquat'. So the rule is: look up the ingredient,
not the brand name.

2
The History of Pesticides

The pesticides tale is usually told like this. They developed in two phases. The first phase was one of haphazard discovery. The second phase, beginning around the time of the Second World War, saw pesticide development based upon scientific discoveries.

While containing a germ of truth, this version of history glosses over the facts. The real pesticides story is one of people and companies using whatever materials were to hand, trying them on whatever pests were around, and seeing which were effective. It is a story of people with many motives – fame, money, power. It is a story of companies doing well out of war, and of people being paralysed in the Prohibition period in the USA.

The pesticides story is part of a centuries-old attempt to control nature with a fairly heavy hammer. It is a search for practical and technical solutions for pest or disease problems. The science followed the practice, more than the other way round.

Nothing new?

Ever since agriculture began, chemicals have been thrown at pests to ensure that people's efforts were not wasted, that they had good crops, and that they did not starve. An Egyptian papyrus of 1500 BC records formulae for the preparation of insecticides against lice, fleas, and wasps. Homer, the Greek poet and observer of farming, referred around 1000 BC to a pest-averting sulphur. Some eight hundred years later the Roman writer Cato advised boiling a mixture of bitumen so that the fumes would blow through grape leaves in vineyards and rid them of insects.

Gardeners and farmers in the past felt that the worse the drug smelt or tasted, the more good it must be doing. One remedy used in England in 1711 for cantharid flies involved 'pouring or throwing on the tops of trees, by means of a small pump, water in which has been boiled rue [a herb with a strong smell and bitter taste]'.[1] Early in the seventeenth century John Parkinson wrote a celebrated book called *Paradisus, The Ordering of the Orchard*. He recommended the use of vinegar to prevent canker on trees:

The canker is a shrewd disease when it happeneth to a tree; for it will eate the barke round, and so kill the very heart in a little space. It must be looked into in time before it hath runne too farre; most men doe wholly cut away as much as is fretted with canker, and then dresse it wet it with vinegar or cowes pisse, or cowes dung and urine, &c untill it be destroyed.[1]

In 1763 there appeared in Marseilles a remedy for plant lice. A tin syringe with a nozzle pierced by a thousand holes was filled with a mixture of water, slaked lime and bad tobacco. In 1791 'Forsyth's Composition', named after its inventor, was used to help heal wounds in trees: 'Take one bushel of cow dung, one half bushel of lime rubbish from old buildings, one half bushel wood ashes, one sixteenth bushel pit sand.' By 1797 experiments were being conducted such as burning brimstone under trees, piling ashes round the roots, and throwing powdered quicklime or soot over the trees when wet, to see which was most effective.

The 1800s saw many developments. In 1821 the London Horticultural Society heard how sulphur was the only specific remedy for the treatment of mildew on peaches. Mixed with soapsuds it could be dashed violently against trees by means of a rose syringe. William Cobbett, the author of *Rural Rides*, gives instructions for the treatment of cotton blight (woolly aphis) which would dislodge the pest. 'Wash the place well with something strong, such as tobacco juice. The potato very nearly poisons the water in which it is boiled – that water would cure cotton blight. Rubbing the part with mercurial will certainly do it.' Thomas Gardner gave an interesting list of current remedies in 1832:

Insects may be annoyed and oftentimes their complete destruction effected by sprinkling over them, by means of a syringe, watering pot, or garden engine, simple water, soap suds, tobacco water, decoctions of elder – especially the dwarf kind – of walnut leaves, solutions of pot and pearl ashes, water impregnated with salt, tar, turpentine, etc: or they may be dusted with sulphur, quicklime, and other acrid substances.[2]

One commentator of the time remarked: 'With such a battery

of powerful materials directed against them it is a wonder that so many insects we have to contend with should still exist. The very number of the materials named is an indication of weakness.' Perhaps somebody should make the same comment today.

The first mass use – Colorado beetle and grape mildew

In 1867 farmers in the United States were faced with an invasion of Colorado beetle, a pest of potatoes, into their crops. They applied an arsenical poison known as Paris Green. This intervention was the foundation of chemical formulation, the beginning of modern pesticide use.

The word 'syringing' was replaced by the word 'spraying' towards the end of the century. The two words mean much the same thing. A chemical is thrown at the plant as a fluid or semi-fluid, in the form of a fine rain or mist. Spraying refers to the operation out of doors, and coincided with one of the biggest steps in pesticide history.

In 1882 a French scientist, Professor Millardet, was strolling through the vineyards in Médoc in the Bordeaux region. There had been a lot of mildew that year, so he was surprised to see that the vines next to the path were still in leaf, while all the rest were bare. He stopped to look at them and found a bluish white deposit on them, as if they had received some form of chemical treatment.

Millardet spoke to the vineyard manager, who said his workers splashed a poisonous-looking substance on to the vines next to the path to discourage passers-by from pilfering the grapes. It consisted of a mixture of copper sulphate and lime. The manager had not thought about it before, but now that the professor had mentioned it he would do some trials next year. He would try anything in the hope of getting rid of the mildew which ruined his grapes.

The professor recognised what he and others had been looking for – a vulnerable stage in the life of one of the fungi that killed plants. (There was still no control for potato blight, despite its ravages since 1845. Millions had died from poverty and starvation in Ireland.) Millardet could see that the chemical mixture had protected the vine leaf from the fungal spore. Unfortunately for him, the next year was not a bad year for mildew. They had to wait till 1884, hoping the mildew would be severe. It wasn't. But it was enough.

You can imagine that much was at stake. Fame would come to whoever could save the vines of France and the potatoes of Ireland. Many other fungus diseases threatened tomatoes, fruit trees and roses. This could be the discovery that had eluded agricultural scientists for years. Rumours of Millardet's success started to reach his scientific rivals in another famous wine-growing district of France, Burgundy. In May 1885 Millardet communicated his precious formula to the Gironde Agriculture Society. It became known as Bordeaux mixture, after the region in France where its value was first recognised.[3]

That year, many growers put Bordeaux mixture to the test; 1885 was a bad year for mildew – except for those growers using the new concoction. It was such a success that all sorts of other people tried to take the credit for mentioning it or having used something similar before. Millardet was so jealous of his discovery that he wrote in the autumn of that year:

'I claim the honour of having been the first to conceive the idea of the treatment with copper; the first to experiment with it, and the first also to recommend its use in practice.' He wanted his fame. He was not bothered about money.

The science of pesticide discovery

Formerly, when a pest injured a plant it was no uncommon practice to apply any remedies or materials that came to hand, regardless of their probable efficiency. It was not generally the weakest point of the organism that was assailed. In many cases it was not even the proper organism which was held responsible for the injury. Nevertheless many valuable discoveries came from these varied and desultory treatments, and some of the remedies most highly prized today were discovered merely by chance, not very many years ago. One by one the enemies have been carefully studied. Only by understanding them thoroughly can proper steps be taken to check their ravages. Yet it is only within comparatively recent years that this first step was taken.

This was written by someone called E. Lodeman, as far back as 1896.[4]

The idea of selective weedkillers – chemicals which could kill certain plants (weeds) and leave others (the main crop) alone – came again from the French vineyards. A vine grower called

Bonnet noticed in 1897 that the plant charlock was killed by a copper sulphate solution, whereas oats were not touched. In the USA a few years later a researcher called Bolley noticed a similar effect caused by ferrous sulphate, which was used for many years afterwards to control broad-leaved weeds in cereal crops.

The first synthetic insecticide, potassium dinitro-2-cresylate, was marketed in Germany in 1892. Thus the three main groups of pesticides – insecticides, fungicides and herbicides – had all been discovered by the turn of the century. And luck, not for the first time, had been accorded a scientific status.

The industry takes off

During the early years of the twentieth century mainly inorganic substances were introduced. These were things such as sulphur derivatives, arsenicals and compounds of lead, copper and mercury. In addition a few natural substances such as nicotine, pyrethrum, derris and quassia, together with crude distillates from tar, were added to the pesticide armoury.

There were attempts to make use of products and by-products from the dyestuff and pharmaceutical industry. In 1932 a rapid knock-down agent for the control of household pests was marketed. It had the delightful name of Lethane.

The Second World War stimulated three major developments in pesticides. The three key discoveries were of

● the insecticide DDT
● the organophosphorus insecticides
● and the selective phenoxyacetic herbicides such as 2,4-D and MCPA.[5]

Despite their complicated names, their stories are worth retelling.

DDT

In the war, armies moved into mosquito- and fly-ridden areas. The need to protect the soldiers and control these disease-carrying insects became urgent. Paul Muller, working for the J. Geigy company in Switzerland, found that a chemical that had been manufactured sixty-five years before and left untouched – and was still stable – had insecticidal properties. Samples were

sent immediately and secretly to the USA and UK, where
further tests were carried out.

For days volunteers wore underclothes treated with this
chemical – now known as DDT. The chemical was still effective
after three weeks. There were no signs of acute poisoning, but
there was a rumour that it had an adverse effect on male virility.
After these tests DDT was used in 1944 to control a major outbreak of typhus in Naples.

Following this discovery other organochlorine compounds,
as they were called, were examined and synthesised. One of
these, benzene hexachloride (now known as Lindane), had first
been prepared in 1825 by the UK scientist Faraday. Its insecticidal properties were not discovered until British (ICI) and
French workers looked at it again following the DDT discovery.

Organophosphates and nerve gases

The first indications that organophosphates were poisonous did
not come from scientific research on other pesticides, but from
bootleg liquor in the 1920s' US Prohibition period. The addition of the chemical tricresylphosphate to make innocuous
liquids into a substitute for alcohol was a deadly recipe. This
gruesome real-life experiment resulted in many cases of limb
paralysis and even death. But organophosphates quickly became
hot news to pesticide chemists. It's an ill wind. . . .

Research into the possible use of organophosphate compounds
as nerve gases was begun in laboratories in Britain, France,
Germany and the USA at the start of the Second World War.
Some are among the most potent nerve gases known. By 1945
in Germany IG Farben had produced 12,000 tons of the
organophosphate nerve gas Tabun. Of their two factories, one
was bombed before being occupied by Russian forces and the
other dismantled and taken to Russia. Old nerve gas shells were
found in the lake at Spandau where all nerve gases were tested
by the military laboratory group known as Weapons Testing
Group 9. The British army was given the responsibility of disposing of them. However that decontamination was still going
on in 1990 under the supervision of the Chief Forensic Scientist of the Berlin Police, Dr Spyra.[6]

Also in Germany, Gerhard Schrader patented organophosphates in 1937. He found one that was not as toxic to humans

but very potent to insects. Schrader's employer, Bayer (previously IG Farben), produced the first organophosphate, Parathion, for crop protection. Others, including Schradan and Demeton, followed in rapid succession. It is not clear what the relation was between the military and Schrader.

Selective herbicides

The third major group of pesticides developed during the war – the phenoxyacetic herbicides – were also not developed as a result of research aimed at assisting horticulture. According to a manager in the Research Department of ICI, they were discovered as part of a secret wartime project to develop compounds which could destroy ricefields and other vegetation. The idea was that they might be used as a weapon to starve enemy populations into submission.[7]

In the early 1940s scientists at the ICI Research Station sprayed cereals with a hormone known to control vital growth processes. The weeds were killed but the cereals remained unaffected. The scientists conducted further experiments and found that a related compound known as MCPA was the most effective. At the same time scientists at the Government Research Station at Rothamsted reached the same conclusion by a different route. Studying a similar chemical (IAA), they realised they were looking at a potential weedkiller. They too looked for more effective but similar chemicals, and came up with 2,4-D. MCPA and 2,4-D are still among the most widely used herbicides today. The two original papers outlining their discovery appeared in the same issue of the journal *Nature* in 1945. Their publication had been held up until this date for security reasons. 2,4-D was later used in Burma for 'anti-insurgency measures'.

The postwar manufacturing and farming revolution

From the end of the Second World War, the farming industry intensified production. Pressures to grow more, using less labour and more machinery, relied upon increasing state funding and agrochemicals. The main pesticides were developed in terms of both chemical structures and manufacturing output. Other members of the organochlorines were discovered in the late 1940s. Chlordane was the first of a series made using the so-called 'diene' synthesis. This reaction was discovered by two chemists, Diels and Alder, who gave their names to the most important members of the group – Dieldrin and Aldrin. It has been estimated that more than a million different organophosphate compounds have been made and tested as insecticides. Of these about 50 are commercially viable.

Another range of compounds, chemically different but with a similar killing action to organophosphates, are the carbamates. These were introduced by Geigy (now Ciba Geigy) during the 1950s. Further phenoxy acids (including 245-T) were developed during the 1950s. The carbamate family grew with phenoxybutyric compounds (MCPB), phenoxy-proprionic acids (Mecroprop), benzoic and phenylacetic acids (236 TBA), all developed to improve the selectivity of weed control.

You can find out more about the general growth of pesticides since the Second World War in Chapters 7 and 8.

Chemicals in search of a use

So what have we learned from this Cook's Tour of pesticide history? Obviously, scientific development is not quite the simple process of men working in white coats in laboratories and suddenly shouting 'Eureka!' There has, however, been a lot of hard work. As you can see, there are three main ways in which pesticides have been developed.

In the process known as random or speculative screening, every chemical found or synthesised for whatever purpose is tested for any activity against any pests. For instance, aminotriazole was used in photography before becoming a herbicide. Pat Mooney of the International Coalition for Development Action put it like this:

If a chemical won't cure cancer, it might wax a car, de-louse a cat, froth a beer or rid a wheatfield of leafy splurge. Sometimes a company can strike lucky. Ethylene dibromide's (EDB) first use is as a scavenger in leaded petrol. EDB also serves as an intermediate in the pharmaceutical industry, acts as a solvent for gums and waxes and does increasing duty as a fumigant for grain and fruit.[8]

Once such a chemical has been found, then the second stage

of development can occur. Researchers look into closely related compounds and try to work out similar combinations. Following the success of DDT and Lindane, many companies developed the organochlorine group.

The third line of development is for researchers to study the biochemistry or life history of the pest. They try to formulate chemicals that will interfere with the pest at crucial points. This was how phenoxyacetic acids were developed.

The myth sold to agricultural science students (two of your authors were) is that pesticide development has come from steadfast scientific research, led by the humane goal of improving food production. Reality, as this history has shown, is messier, sometimes pretty unsavoury and dominated by companies looking to find a use for compounds that their laboratory workers have created or discovered.

Almost every breakthrough has come from the 'hit and miss' method. Those that have not have come as a side-effect of war. Bear that in mind next time you see an agrochemical company advertisement implying that if the public does not support their products, everyone will starve, or the Third World will not get clean water. Who is rewriting history?

Where Do You Find Pesticides?

Every year around a billion gallons of pesticide-based liquid is sprayed in Britain. The vast majority is used in farming. In the USA 72% of pesticides go to farming. We estimate about the same for the UK.[1] How and where can pesticides affect you? Everywhere from food and water to clothes, pavements and houses. We outlined some of the likely places in the Introduction. Although we have said that our intention is to inform rather than to scaremonger, some of the following facts may well startle and even horrify you. But they are all true.

- In an average year, over 900 people are discharged from hospital after being treated for pesticide poisoning.[2]

In your food

People have been poisoned by pesticides in food for years.

- 159 people in Wales in the 1950s were poisoned by Endrin that had contaminated flour used for making bread[3]
- In Mexico in 1967, sugar and flour contaminated with Parathion caused 16 deaths and many illnesses.[4]

The UK Association of Public Analysts' 1983 survey of pesticide residues found that:

- just over a third of fruit and vegetables contained detectable residues
- some foods contained residues of pesticides which were not supposed to be used on them
- 13 out of 32 lettuces carried Lindane. This is a persistent organochlorine and is banned in at least 15 countries.[5]

Some food companies set their own standards. In the mid 1980s Heinz USA said that it was banning 12 pesticides from use on its supplying farms. Heinz UK said it had already banned three pesticides – Captafol, Dinocap and Daminozide.

By 1990 government tests had found low levels of residues in:[6]

- 10% of bran-based breakfast cereals
- 55% of wheatgerm
- 93% of pure bran
- 16% of processed oats
- 24% of rice
- a third of sausages sampled
- nearly half of 150 burgers and 177 cheeses sampled
- in 37 out of 46 samples of apples.

In the air

If you live in the countryside or your council sprays your pavements in town, it can be hard to see the drift and to avoid the spray. Every year people suffer from these effects, and some people complain (see Chapter 14). Statistics rely on investigating complaints and outbreaks of mass poisonings.

- in the UK, in the three years to 1989 on average 125 people each year were investigated following complaints. Fifty-five of these were confirmed or considered likely to be the result of pesticides. The pesticides implicated most are Carbendazim, Demeton-s-methyl sulphone, Fenpropimorph, Mancozeb, Maneb and Paraquat. Look these up in Tables 3 and 4 to find out more[7]
- in 1986 Friends of the Earth, through its local groups network, did a special survey on spraydrift. Spraydrift occurs when tiny droplets of pesticide float away from the crop it is meant for. A total of 113 pesticide incidents were recorded: 55% of these referred to problems with spraydrift; 29% were specifically associated with aerial spraying[8]
- about 0.23% of pesticides in use in Britain in 1989 were applied by aircraft.[9] Aerial spraying is the cause of more complaints and accidents than any other form of pesticide application. Only certain pesticides are approved for aerial spraying. They include some known to be hazardous for health: Captan, Benomyl, Chlorpyrifos, 2,4-D, Dichlorvos, Malathion, Metaldehyde and 2,4,5-T
- at least 11 pesticides cleared for use by aircraft are suspected of causing cancer or genetic and birth defects in animals[10]

- aerial spraying is very inefficient. Losses from the target area range from 20 to over 50%[11]
- drift from spraying can be detected up to 500 metres away downwind.[12]

In water

- between 1985 and 1987 water authorities recorded that the limits for pesticide residues in public water were broken about three hundred times in England and Wales. Most breaches were in the Thames, Anglian, Severn-Trent and Wessex areas, a survey by Friends of the Earth found[13]
- Atrazine and Simazine were the pesticides most implicated in the breaches. These are Triazine pesticides which are water-soluble. Atrazine alone accounted for over half of the incidents. One sample in the Thames region found Atrazine at 45 times the European Community permitted limit[14]
- a survey of tapwater in London found that two-thirds of samples contained pesticide levels above both European and UK government guidelines[15]
- in the USA a recent review of pesticide contamination of ground water reported that 74 different pesticides had been found in the water of 38 states. Forty-six of these pesticide detections were due to *normal* use. The most commonly found were Aldicarb, an insecticide; Atrazine, a herbicide; and EDB, DCP and DBCP, all soil fumigants.

At work

- In the three years 1987–9 an average of over 24 farmworkers each year have been confirmed as having been poisoned by pesticides
- Four workers from one Deryshire farm were reported in the *British Medical Journal* in the late 1960s to be sexually impotent. The doctors considered, as there was not any other common factor, that it must have been pesticides that caused the 'Derby Droop'[16]
- In a trade union survey in 1986, 50% of those replying reported classic symptoms of pesticide poisoning – despite 80% of them wearing protective equipment[17]

- a survey published by the Union of Construction, Allied Trades and Technicians in 1988 found that 46 out of 116 respondents had supposed ill-health, which they attributed to wood preservatives[18]
- dermatitis has been reported among workers making many pesticides, including Pyrethrum and Paraquat
- chloracne, a more serious skin condition, has been found among workers making many organochlorine pesticides
- many other workers come into contact with pesticides used to kill creatures such as cockroaches, fleas and rodents that may flourish at work but would never appear at home
- other workers deal with pesticides that are being impregnated into other products – paper and textiles.

At home

- 5000 people in England and Wales suffered acute pesticide poisoning, the National Poisons Unit estimated in 1986[19] of which 69% of enquiries about agrochemicals to the National Poisons Unit resulted from garden or home incidents[20]
- in 1982–3 there were 93 accidents involving pesticides in the home. Seventy-nine of these involved children under 5 years old[21]
- pesticides are used to treat wood rots and moulds. DIY shops sell them openly to the public. About 600,000 houses a year are treated with some form of wood pest control, according to manufacturers Schering UK[22]
- in 1989, the London Hazards Centre showed that half a million people in the UK would in any one year get their largest dose of pesticides just through wood preservatives[23]
- Pentachlorophenol, one wood preservative, is rated as highly hazardous by the World Health Organisation. One safety journal held it responsible for a thousand deaths worldwide[24]
- the UK Building Research Establishment calculated in 1983 that, up to one month after being used, residues of Pentachlorophenol were about three times higher than its own recommended 'acceptable air concentrations'[25]
- a study of bats – insect-eaters and therefore one of nature's own pesticides – found that they died more quickly if they roosted in contact with timbers treated up to 14 months previously[26]

- the London Hazards Centre found that, out of several hundred cases, eight people are recorded as having suffered epileptic fits after timber treatment of their homes[27]
- the Laboratory of Toxicology at the University of Antwerp looked into chronic poisoning in people exposed to wood preservatives containing Lindane and PCP. Of 40 people exposed:

    nearly all . . . complained of constant weakness and dizziness. . . . More than half complained of abdominal pains (sometimes accompanied by diarrhoea) or of an acne-like skin rash, together with a fierce itch and sometimes with growths. . . . One-third of them had chest pains, or a sudden weight loss. Seven people had blood in their urine without any readily available medical cause

- in Vancouver, Canada, two small boys nearly died from poisoning by Parathion found in their cotton flannelette sheets. Prompt detective work by the medical authorities found 140 contaminated sheets, and they stopped shops selling the sheets so that no one else was affected.[28]

In your clothes

There have been some dreadful cases of clothes getting contaminated by pesticides. For instance, in the 1960s a consignment of boys' jeans in the United States became contaminated with Phosdrin. Even though the jeans were then kept in an air-conditioned warehouse for eight months, six children were poisoned before the source of contamination was tracked down and the jeans withdrawn.[29] One renowned toxicologist has said: 'Contamination of clothing and other fabric by pesticides is a potential source of serious injury although contamination of food has caused more illness and death.'[30] You probably buy cotton clothes without any thought that the plant from which the cotton fibres are harvested has been sprayed with pesticides. By the time you wear the clothing, the residues will be almost non-existent. But think about the workers on the farm where the cotton has been grown.

- in the United States, there are over 45 active pesticide ingredients approved for use on cotton. Seven of these were identified by the US Environmental Protection Agency as causing benign or malignant tumours. They were defined as 'potentially oncogenic' (see Chapter 6)[31]

- one study in the United States compared the health of communities living near cotton fields with those in non-cotton-growing areas. Symptoms 60–100% more likely to be found in the cotton-growing areas included: fatigue, eye irritation, rhinitis, throat irritation, nausea and diarrhoea. These problems could not be put down to other effects. The link between pesticide use and ill health was statistically strong, even when other ill health factors were taken into account[32]
- in Brazil, the cotton boll weevil and pink bollworm are the major insect pests of upland cotton areas. Most pest control is carried out by hand-operated sprayers, with the liquid stored on the farmworker's back. Less than 40% of pesticides applied in this way reach their target.[33] Where does the rest go, we ask?

In your garden

Your garden is a mini-farm, and farms today rely heavily on pesticides.

- in 1989 British gardeners spent a total of £23.4 million on pesticides, £9 million on herbicides, £7.2 million on insecticides, £1.2 million on fungicides and £5.1 million on herbicide/fertiliser mixtures[34]
- greenhouses are enclosed places which form ideal breeding grounds for disease. As a result domestic greenhouse owners tend to be sold the advantages of pesticides. A greenhouse-grown tomato, for instance, may have been sprayed many times. According to the US National Research Council, the pesticides used for tomatoes carry the greatest potential for causing cancer[35]
- the British Agrochemical Association says that products for use by amateur gardeners are formulated so that they can be applied safely without the need for protective clothing. There is not, and never has been, anybody representing gardeners on any safety approval committee (see Chapter 11)
- Permethrin is a commonly used garden pesticide. It is approved for use in the UK. Why not look it up in the tables at the end of the book to see what its effects are
- Amitrole has for a long time been suspected of causing cancer. The UK government's Advisory Committee on

Pesticide accepts that it causes goitre, an enlarged thyroid gland. Yet many garden chemicals contain amitrole without any warning on the packet that there may be long-term effects.

As used by your local council

A lot of pesticides are used by local authorities. A report produced by Norwich City Council showed how widely they are used.[36] The sprayers may be council workers or outside contractors. They may be dressed like someone from Mars, or stripped to the waist. The directions for use may say that the operatives should wear protective clothing. But it is easier to make the rules on some committee than to turn up on a hot afternoon and wear all the gear. Often, too, the knapsacks holding the pesticide are poorly designed and leak.

Pesticides can be used for:

- weed control on footpaths; this is the single largest outlet for amenity-use pesticides in the UK. Common weedkillers include Amitrole and Atrazine
- turfcare in parks and sports grounds; on bowling greens moss-killing agents like Dichlorphen are used
- slugs; Metaldehyde is used in slug pellets. There have also been alleged cases of cats and dogs being killed by Metaldehyde
- road margins; MCPA has been commonly used. Green arguments are now being used to rationalise cutbacks on herbicide use for roadsides – an ironical improvement
- council buildings; to protect outside paint and external walls, and for mould treatment inside buildings and in cupboards and on interior walls. One council's records show that in one year it treated 200 full roofs, excluding private contract works.[37]
- rats; in London, rats are now estimated to be almost as numerous as people. Rodent control by councils is common in cafés, restaurants and hotels
- fleas; while the human flea has all but died out, cat fleas have increased dramatically in the last few years with a series of warm summers and winters. In the South-east particularly, there can be waiting lists of several weeks for house decontamination by local councils.

Don't forget the sheep

There are around 40 million sheep in the UK. In 1990 they were dipped on around 19,000 farms. The European Commission is concerned that, after the sheep have been dipped, the leftover dip is not being properly disposed of. MAFF advises farmers to tip the dip into a soakaway or to spread it on fields away from surface waters. The European Commission says that this breaks the 1980 Ground Water Directive, formulated to stop pollution getting to deep water. The Commission says the dip should be incinerated or dumped at licensed landfill sites.[38] In 1985, of 1,492 samples of sheep's kidney fat, 71% contained residues of Lindane.[39] Finding this much Lindane led to a voluntary ban on Lindane in sheep dips. Now farmers are encouraged to use organophosphate dips. But . . .

- in 1987, 7% of 287 samples of sheep's kidney fat contained residues of Diazinon. About 2% exceeded official limits[40]
- 24 out of 25 sheep dips currently approved by MAFF have been sent for review (see Chapter 9 for how pesticides are reviewed)[41]
- Dr Routledge of Newcastle College found at certain times of the year sheep dip residues up to 20 times the World Health Organisation maximum level permitted in drinking water.[42]
- Following this, the Health and Safety Executive (HSE: the UK national organisation that policies safety standards, particularly in working conditions) announced a study of the side-effects of sheep dips. The study was carried out on only 22 people in the county of Hereford and Worcester. Why did they not do it in 'real' sheep country?[43]

And there's more . . .

We have not even touched on other areas where pesticides are in common use, such as

- schools, where any of five pesticides – Lindane, Carbaryl, Permethrin, Malathion and Phenothrin – could be used to combat the increasing scourge of head lice
- railway tracks, where many herbicides are sprayed. In a famous study of Swedish railway workers, an increase of soft cell cancers was found among those spraying a mixture of

pesticides. In June 1990, the Minister for the Environment of the southern German state of Baden-Württemberg set up a new four-year programme of alternative controls for the railways in the region[44]

● transport, about which a UN reviewer concluded over 20 years ago 'from conversations with warehousemen and drivers and from personal experience with spillage incidents that there is a tendency to try to 'clean up' any spillage without contacting the shipper or pesticide company or obtaining other expert advice. The result is inefficient decontamination. . . .'[45]

● shops, where you will find dozens of bottles, cans, packets and tubes with strange-sounding names.

You can also come into contact with the *same* pesticide in a variety of ways. Consider Lindane. You can

● eat small amounts in lettuce
● breathe it in after wood preservation at home
● drink minute residues, in water, from sheep dips
● get it in your hair.

You can also find it in situations ranging from chocolate bars to the insides of emergency vehicles – ambulance workers have been contaminated with it when fumigating their vehicles. Nobody has investigated the possible cumulative, multiplying or sensitising effects of each of these exposures. (A cumulative effect means the effects get added, the more pesticides are used. A multiplying effect means the presence of one pesticide magnifies the effect of another pesticide. Sensitising is when a minute dose of the pesticide starts to have the same effect as larger doses used to.)

4
What Do the Labels Mean?

When you buy a pesticide it must have a label – by law. Whatever bottle, packet, spray or can of pesticide you buy, you will find a label.

In the UK a major part of the approval process involves drawing up the label. Pesticides are considered safe only if used according to the directions, so we need to read and understand the label. It is estimated in the USA that only 4% of farmers read the labels on their pesticides.[1]

What the label should say

Pesticide labels must conform to certain rules set out in the Data Requirements for the Control of Pesticides Regulations (COPR) 1986. This law incorporates the Classification, Packaging and Labelling of Dangerous Substances Regulations 1984, which includes various EC directives.

The guidelines in COPR 1986 are almost identical to those of the previous voluntary Pesticides Safety Precaution Scheme (PSPS). There is one interesting exception. Under the old PSPS system the label 'must *not* claim that the product is "safe", "non-toxic" or "harmless" in terms of risks to humans or animals'. No such qualification now exists!

By law, the label

- should be clearly set out and contain sufficient information to ensure safe handling
- should be resilient to wear
- must be in a prominent position
- must include the common name (according to BSI, the British Standards Institute) of the active ingredient.

This is why we have used the common names for pesticides in this book. Read the label carefully – the name often appears in small print. From the common name you can check out the pesticide in this book. You can also look it up in a book called

Pesticides published each year by HMSO. This contains the list of MAFF- and HSE-approved pesticides. If the pesticide is not there, it is not approved.

Every label must include the following:

Label	What it Means
Wondersplat	Trade name
Contains 1% W/W Diatronprep	Common name with concentration
For use only as twigicide	A phrase restricting use
How to use	Manufacturer's recommendations on use, e.g. rate of application
Precautions, e.g. keep away from children	The law requires these instructions to be followed
Chemco., Tweedsdale	Name and address of holder of approval
MAFF 1 2 3 4 5	Registration no. of approval

Dangerous pesticides also have to show

- the amount of active ingredient
- a clear 'risk' phrase
- a symbol appropriate to their danger
- and a way to identify the batch.

Dangerous pesticides are those classed as:

- very toxic, i.e. with an oral rat LD50 of 5 mg/kg or less
- toxic, i.e. LD50 of 5–50 mg/kg
- harmful, i.e. LD50 of 50–500 mg/kg
- explosive, i.e. those exploding under flames
- highly flammable, i.e. those substances which could catch fire under normal conditions
- flammable, i.e. liquids with a flash point of 21°C or over
- oxidising, i.e. substances which release heat on contact with other substances
- corrosive or irritant, as defined by authorities such as the Ministry of Agriculture or the World Health Organisation authorities.

'LD50' means the lethal dose for 50% of a laboratory sample of animals. So an LD50 of 5 mg/kg means that half the animals die when they eat 5 mg of poison for every kilo that the animal weighs. This is a very crude assessment of how poisonous a pesticide is, but a very common one. The value of LD50 tests is discussed in Chapter 9.

Finally, the label should also show directions for use, or how to obtain information for use.

What else the label could say

Much depends on the labels being read and understood and the instructions followed. The UK authorities presume that, because the words are on the labels, everything is all right. Yet look at one phrase that appears on the label of certain pesticides. '. . . is an organophosphate/anticholinesterase compound. *Do not use* if under medical advice *not* to work with such compounds'.[2] Can you get a grip of that? Has anybody ever gone to the doctor and asked: 'Can I work with an anticholinewhatsit?'

We believe that the label should also say something about reducing exposure. At present all the emphasis is on you wearing protective clothing. Yet under the Control of Substances Hazardous to Health (COSHH) regulations the law requires the removal of the hazard at source by using engineering or better controls to prevent the hazard. This is to be the first line of defence – not the last. As labels are the important point of contact they should state that *reducing exposure* is the first thing to consider. They could tell you to substitute, suppress, enclose or ventilate the chemical.

The label should also carry symptoms of poisoning. Then you would know what to look out for! Symptoms are given in Table 4.

We say that:

- labels are necessary, but not to be relied upon
- there should be more information either on the label or linked to it
- proper training and education on what the labels mean should be given to everybody, not just those few covered by present laws.

Data sheets

Labels should be linked to data sheets. Pesticide makers produce but do not publicise these data sheets. Get them. They contain more details than the labels, and can help you assess whether the particular chemical has to be used. Information in data sheets should include:

- chemical name. This is useful if you want to look into the medical or technical literature
- toxicity data -- e.g. animal LD50 studies
- symptoms of poisoning by inhalation, ingestion and skin contact
- emergency procedures, e.g. for fire, first aid and spillage, and advice to doctors
- storage recommendations
- ecotoxicity, i.e. what they can do to the environment
- supply and transport labelling data.

Data sheets could be improved and standardised. Some are good, some are very brief, and some are too technical. Some are actually misleading. Get hold of the data sheet for Bendiocarb under the trade name 'Garvox 3G'. It talks about it being a 'direct inhibitor of cholinesterase' and causing 'brachycardia'. Under 'Fire' it says: 'Do not breathe dust or fumes. In common with all other methyl carbamates, bendiocarb will liberate strongly lachrymatory methyl isocyanate (MIC) when heated above its decomposition temperature.'[3] You may be able to work out that you could end up in tears. But few would know that it is the same MIC that was released from Bophal in India, killing thousands! 'Tears to your eyes' must rank as the understatement of all time.

Labelling residues in food

Pesticide residues in foods are not labelled at all. You will know that some pesticides end up in food, but you cannot see if your food has been treated with pesticides. A label could say, for instance: 'Treated with Aldrin'.

Since 1984 the European Commission has made all food manufacturers declare food additives on the product label.[4] So you can read a label and see something like this: '*Ingredients*: polyphosphate (E450), colour (E-102), antioxidant (E301).' If you know that E-102 is tartrazine and want to avoid it, you don't buy that food.

The first consequence, after this became law, was that everyone discovered additives going into their food which previously they hadn't known about. Secondly, some people did some home detective work. They began to note that when they or their child ate a particular food, behaviour changes or allergies followed. It was so simple to arrive at this conclusion that some food manufacturers started playing cat and mouse with consumers, and took away the E numbers and replaced them with their long chemical names.

So why can we not have food labels saying which pesticides have been used in growing or making the food?

Changes are, however, afoot in Europe which may help consumers. There are proposals in the European Pesticide Directive to set up a Brussels-based central registration scheme for new active ingredients for pesticides. Earlier attempts in the 1970s and early 1980s to harmonise a Europe-wide system of pesticide approval foundered in 1983.[5] But with the advent of the Single European Market companies are now more positive about European harmonisation – for trade purposes – and resistance to such a scheme seems to have evaporated.

If there can be a central database on new pesticide ingredients, we cannot see why you, the consumer, could not benefit as well. This would mean:

- open access to the information on the Brussels database
- a label being attached to food products. It could tell you which ingredient has been sprayed on the food.

Post-harvest treatment labelling

So how likely are we to get pesticide residue labelling? There is good and bad news.

In 1989 there began to be talk in European circles of introducing a new label. The one being talked about was this: 'Treated with post-harvest chemicals'. Various authorities in Germany, the Netherlands, France and Sweden have expressed concern about the safety of post-harvest treatments.[6] In Germany and France, citrus fruits that have received post-harvest pesticide treatment have to declare the details on the label. Sweden has gone further and virtually banned such treatments

except for some long-stored potatoes. That country considers
that post-harvest residues are too high and that the treatment is
not necessary: alternatives such as cold or chilled storage exist.
In 1990, the European Commission rejected even this proposal
to have post-harvest treatment labelling (see p. 93).

Post-harvest treatment residues are usually substantially
higher than pre-harvest pesticide residues. We want Europe to
have a system not just for post-harvest labelling but for all treatment labelling.

- many pesticides on food are applied on the farm – before
  harvest
- a post-harvest treatment label would understate pesticide use
- labelling just for post-harvest treatment implies that unla-
  belled foods have no residues
- the 'treated with post-harvest chemicals' wording is too vague.
  It doesn't tell you which pesticide has left residues on the food.

Other desirable forms of labelling

Say you have received a request from the Norwegian environmental movement to avoid all chlorines. Chlorines are the base
of some bleaches, plastics and some common pesticides. Or
you may be allergic to Lindane. Whichever, you may wish to
avoid food with possible contamination from organochlorine
pesticides. For you the 'post-harvest treatment' label would be
useless.

Pesticide producers and food manufacturers are nervous.

- they don't want another wave of consumer concern like the
  one that arose with the E numbers. Yet they also tell us we
  have a choice. How can we choose if we do not know what to
  choose from? If a pesticide is used, why not say so?
- they also say that the labels would be very complex. We
  agree that it would be complex if they had to label the
  amount of residues. Take bread. A standard white sliced loaf
  is made from wheat from many different farms and countries, depending on the time of year. It would be impossible
  to test each loaf batch for residues and then label the bread
  accurately.
- It would, however, be possible to show that a pesticide had
  been used. Retailers already know what treatments the food
  that they sell has received on the farm. The new Food Safety

Act 1990 has introduced the 'due diligence' clause. This demands that food providers make all efforts they can to ensure that their food is safe and good for health.[7] If these providers are having to double-check, why do they not state what has been done to their food?

We want a label containing:

- what pesticide has been used on the food
- each pesticide named and numbered
- standard numbering for the whole of Europe.

It would be based on the additives labelling system for European consistency, and could look like this:

'This food was grown or processed using the following chemical pesticide: Mancozeb (P–557).'

In 1992 a new, expanded 'E' additive numbering scheme starts: a good time to do likewise for pesticides. A 'P' for Pesticides numbering scheme would give consumers a real advance. You have been told how important for business is the removal of barriers to trade by the end of 1992. Why should there be any barriers for *you* to trade in 1992 as well? Here's to a simple P for Pesticides labelling scheme right across Europe!

What's in Your Food, and Who Tests It?

Pesticides can end up in your food in a variety of ways. Residues might be due to:

● accidental spillage leading to possible poisoning – sometimes mass poisoning
● contamination from spraying.

Mass poisonings

Here are some serious cases.

Case no. 1. The worst case we have come across took place in Iraq. In 1971, wheat and barley seed which had been treated with a methyl mercury fungicide was eaten by humans. The warnings not to eat the seed but to use it only for planting were in Spanish or English – not much use in Iraq. Thousands died. Official Iraqi figures say that 6,530 people were poisoned and that 459 died. Oxfam estimated that nearly a hundred thousand people were poisoned and that six thousand died.[1]

Case no. 2. On Saturday evening, 29 June 1985, 264 cases of poisoning from eating contaminated Californian watermelons were recorded in Oregon, USA. Sixty-one cases were later confirmed. Two nights later a second incident occurred, and then a third. On the eve of the 4 July Independence Day celebrations all watermelons from one distributor were recalled. It was too late, and cases were reported from several other states. A major scientific inquiry began, and found that Aldicarb was the cause. The poisoning symptoms emerged 30 minutes after eating the melon and included nausea, vomiting, abdominal pains, diarrhoea, blurred vision and excessive salivation.[2]

Case no.3. During May and June 1985, the Vancouver Public Health Board investigated over three hundred reports of illness. The symptoms reported included: nausea, vomiting, dizziness, muscle twitching and blurred vision. At least 140 people were confirmed as having eaten cucumbers adulterated with Aldicarb.[3]

Aldicarb is and was manufactured by the Union Carbide company, and sold under the brand name Temik. This is still on the UK market. In 1977–9 it was found in ground water in Long Island, New York, having been put on many potato crops there. Tougher controls were introduced. But despite these, five or six years later the cucumber incident occurred.

Mass poisonings occur in four types:

- due to consumption of meals made from ingredients contaminated in transport or storage. The worst case involved 691 people and 24 deaths
- due to meals made from ingredients treated with pesticides. Besides the Iraqi case there was another in Turkey where between three and five thousand people were affected by Hexachlorobenzene. Three to 11% of those affected died *each year* for four years after the poisoning outbreak
- accidental poisonings of food. The most serious was in the USA. Sodium fluoride, a wood preservative for industrial use, was added to hospital meals, affecting 263 people; 47 died
- due to misuse. The worst case was the Californian watermelons, in which 1,175 cases were finally recorded.[4]

Continuous contamination

'Most foodstuffs on sale in the UK don't contain any detectable pesticide residues.' So said Chris Major of ICI Agrochemicals in 1990.[5] Various official reports tell more of the story. There are two sources of testing in the UK. Members of the Association of Public Analysts (APA) test on behalf of local authorities. The APA occasionally publishes its own reports. The government's Working Party on Pesticide Residues (WPPR) publishes figures every year. Look at Table 1 for their tests on particular foods. Various studies of theirs in the 1980s showed that:

- many foods, including baby foods, health foods and staples like bread and vegetables, contain residues. Twenty-one % of infant foods, 37% of animal products, 47% of cereal and cereal products and 26% of fruit and vegetables are contaminated

- some UK and imported staple foods such as bread, fruit, cereals and vegetables contained above the legal residue levels. 2% of foods tested overall, 6% of cereal and cereal products and 6% of fruit and vegetables were above the Maximum Residue Levels (MRLs). A MRL is a set level for residues. It is explained in detail in Chapter 10. To exceed a MRL doesn't necessarily mean serious danger, but it is a warning, nevertheless.

- key foods such as meat were not tested
- some foods contained over 30 pesticide residues in all
- Lindane was in no fewer than 11 kinds of fruit and vegetables.[6]

The latest report of the WPPR was produced in 1990 and confirms the earlier tests. Dr Stanley, the WPPR chairperson, said at the press launch of this report that 'there is no such thing as pesticide-free food in Britain'. He added, however, that levels were very low.

Fruit cocktail

Take the apple, that symbol of health and a wholesome diet. You are all being urged to eat more fibre, less fat, and to go for naturally occurring sugar. What could be better than the apple?

In 1990 the Consumers' Association showed that 99% of all apples receive pesticide treatment. Residues decline as time passes. The report found that peeling apples removed 90% of the residues.[7] Is there a problem?

MAFF tests apples for 40 residues, yet more than 70 pesticides can leave residues. Of the four hundred or so pesticides approved for use in the UK in 1989, 135 were passed for use on apples. Nearly 50 are used in practice. Some 20% of the 40,000 tonnes of neat chemicals applied on UK farms is sprayed or used on fruit. Official 1983 figures (the latest available ones) showed that 99% of apples received an average of 26 treatments.[8] About 30 of the 50 pesticides used on apples are known to have caused ill effects in animal tests or to farm workers at high doses. Yet they are approved.[9] The Association of Public Analysts (APA) found nine apples contaminated out of 42 tested. It found five different pesticides. Let's look at these figures in detail:

Pesticide Residues Found in 1983 in Fruit Sold in the UK

fruit	number sampled	number contaminated	number of pesticides
apples	42	9	5
blackcurrants	11	4	2
cherries	11	5	6
grapes	23	6	6
lemons	12	3	3
oranges	20	8	5
pears	37	19	5
plums	15	4	2
raspberries	17	10	5
rhubarb	3	3	2
strawberries	35	14	17
tomatoes	33	11	6

Source: APA 1983.[10]

Levels were measured in milligrams of residue per kilo of produce weight (mg/kg). 1 mg/kg is equivalent to one part per million. So residues are small.

And now for Daminozide . . .

Until 1988 about 10% of the UK apple crop was sprayed with a plant growth regulator called Daminozide. Daminozide is a cosmetic pesticide: it helps plump up the apple and get it looking nice. Daminozide was one of those pesticides which have had a few question marks over them for some time. Few apple eaters took much notice of an obscure safety debate in scientific and academic circles from the late 1970s. The concern was over whether Daminozide was linked to cancer.

But New Zealand did take notice, and in 1985 its Apple and Pear Marketing Board banned Daminozide on its farms. This action showed that good apples could be grown without it. In 1987 the National Research Council of the USA expressed concern about Daminozide. The United States Environmental Protection Agency estimated that Daminozide could cause 45 cases of cancer for every 1 million people eating apples over a lifetime. In 1989 the story broke first in the USA, then worldwide.[11]

The heart of the Daminozide issue was this: when processed into apple pies or apple juice, Daminozide residues turn into a chemical by-product called UDMH. UDMH was the possible danger. A subsidiary question was: why use an unnecessary cosmetic that poses an extra risk?

In October 1990, the government released its 1989 Daminozide and UDMH apple test results. It had found that:

- two out of 55 samples of apple juice had Daminozide in them
- seven out of 22 samples of apple products contained UDMH.[12]

Let's hope the levels are dropping since the product has been withdrawn by the makers. In the USA, the 1989 report by the Natural Resources Defense Council which sparked the public debate has been criticised and is subject to a legal case. MAFF still says that Daminozide is OK, but the National Farmers Union advised growers not to use it.

Vegetables

Twenty % of the vegetables tested by the APA showed pesticide residues at or above the limits of the day.

Vegetables Found Contaminated in 1983

vegetables	number sampled	number contaminated	number of pesticides
Artichokes	1	1	1
Aubergines	3	2	2
Beans	4	1	1
Beetroot	1	1	1
Cabbage	17	2	3
Carrots	4	2	2
Cauliflower	13	3	4
Celery	4	2	2
Courgettes	5	4	3
Cucumbers	16	6	2
Lettuce	32	13	7
Mushrooms	39	12	8

Onions	3	2	1
Peppers	5	1	2
Potatoes	5	5	2
Spring onions	3	1	2
Turnips	2	1	1
Watercress	3	2	3

Source: APA 1983.[13]

The 1986 WPPR report gave test results for 1982–5. It found:[14]

- 43% of 1,649 fruit and vegetable samples had detectable residues; 29% had residues in excess of United Nations levels
- in a number of cabbage and brussels sprout samples, DDT residues exceeded the MRL residue limit. This indicated that DDT was still being used despite having been banned in October 1984
- one lettuce contained 20 times the MRL residue limit for a dithiocarbamate fungicide, the group which is associated with cancer, genetic mutation and birth defects.

Far too little research has been undertaken on how pesticides change when the food is processed. Look at Daminozide and UDMH. Pesticides, being chemicals, can combine with other chemicals in food, to create the so-called metabolite or by-product problem.

Look at what can happen with the potato. In 1991, Birmingham's public analyst found ETU, the metabolite from EBDC fungicides, in crisps above WHO MRLs. In 1990 the Consumers' Association laboratories baked some jacket potatoes. The testers found that 85% of the Tecnazene residues and about half of the Chlorpropham residues disappeared. Good news. But the low level of Tetrachloroanaline a metabolite *went up* by five or six times. Nobody had discovered this before![15] This result demolished the idea that you get rid of pesticides when you cook food. It also questioned the adequacy of government monitoring procedures. Who is looking for breakdown products?

Oils

The 1986 WPPR report found that all the samples of cod and halibut liver oil – the new health products – contained organochlorine residues. They are highly persistent, which means they don't break down easily.

Meat

Animals are basically vacuum cleaners. Nowadays they are given a wide range of foods, which means they eat a wide range of pesticides. And the meat concentrates the residues.

In 1987, the National Academy of Sciences (NAS) in the United States published the results of a study on pesticides it had conducted for the US government. The NAS drew up a list of foods from which there could be a risk. Beef came second and pork eighth.[16] Here is the full list, worst first: tomatoes, beef, potatoes, oranges, lettuce, apples, peaches, pork, wheat, soybeans, beans, carrots, chicken, corn, grapes. Although the calculations were contentious, no comparable study for the UK has been done.

After this concern by the US government about residue dangers from meat we expected some food testing of UK meat. Yet the 1990 WPPR report tested only six samples of UK pork compared to 30 imported samples. Only seven UK beef samples were tested out of 27. Only one UK lamb was tested, yet 69 imported lambs were. And yet there were 29 pheasant and 23 woodpigeons tested.[17] Work out who is protecting whom!

Breastmilk

Humans are animals too. The WPPR report of 1986 found that breastfed babies could have been exceeding the World Health Organisation's limits for DDT through levels contained in their mothers' breastmilk. In some areas of Germany a breastfeeding mother can have her breastmilk tested by state analysts for pesticide residues. It is also possible for women to have their milk tested earlier, but only if they pay for it.

Pressure is growing to give British mothers the right to have their milk tested, as in Germany. The UK Department of Health's Committee on Toxicity reported that between 1980 and 1984 it had tested 40 samples of breastmilk from one region of the UK. It tested only for organochlorines, and the following residues were found: Beta HCH, Dieldrin, DDT, DDE, HCB and PCBs.[18] The WPPR recommended further studies on a national basis. None has yet been done. The 1989 WPPR report says that no follow-up study has been conducted because of shortage of samples. Over a million babies were born in the UK since 1984. We wonder where the committee was looking!

We call on the Department of Health in the UK to test a representative group of mothers for pesticide residues immediately. And they should pick women who live or work in places where pesticides are used. As the British Medical Association BMA 1990 report said: 'Although the presence of possible residues in breastmilk is an area for genuine concern . . . it is important that women should not be led to believe that breast-feeding is in some way less healthy than using formula foods.'[19]

Body fat

The 1986 WPPR report made a small study of organochlorine pesticides in human fat. Organochlorines are persistent – they don't go away. They end up in human fat. In 1982–83 187 samples of human body fat were obtained from people.[20] There has been a decline in general organochlorine levels since tests were last done. Good news – but hardly surprising, as organochlorines are used less these days. The WPPR recommended that further studies should be done, but none has been.[21]

Blood

Testing for residues in blood is quite expensive – over £100 per time. We would like a national survey done on residues in the blood of at risk groups.

Monitoring those residues

Why isn't more testing done? First, testing for pesticide residues in food is an expensive business. It can cost up to £30,000 to test just 30 or so samples for some pesticides.[22] This means that only a small proportion of the money the public spends on food can be used for testing.

Whenever an official report comes out, MAFF says it shows a 'reassuring picture'. Yet the testing done by the Ministry is hardly exhaustive. Their best efforts test for at most 10% of pesticide residues which might be present. Nor does the Ministry recognise that the testing budgets of local authorities are ridiculously low. They can test only one food product for every 100 million bought and consumed in Britain.[23] In 1990, the British spent around £45 billion purchasing food and drink. In the same year, 16 public analysts handled only a thousand samples of food. They spent between £80 and £300 per sample analysed.[24]

In 1985, the London Food Commission asked three major retailers what their testing procedures were. One, Marks and Spencer, said they were confidential. The other two, Tesco and J. Sainsbury, were helpful. At that time, J. Sainsbury did not test regularly. Tesco sent samples to a public analyst to check for the eight most commonly found residues. J. Sainsbury only had tests done by an analyst if they suspected something was wrong – they get their suppliers to have analyses done.[25]

Since that time almost all major UK retailers have taken much more interest in pesticides. At least three have undertaken major rethinks, and all are having to consider their policies very carefully. This is due partly to heightened public awareness, and partly to changes initiated by the Food Safety Act 1990 (see Chapter 13). This piece of legislation makes food companies more cautious about pesticides: it says that companies have to show that they have taken all due care and precautions to make sure that food is safe.[26]

Researching for this book, we wrote to the UK's top 20 retailers asking them what tests they were doing and whether they would publish results. Only two replied, and none would release data. We learned from this that:

● companies are cagey about you or us knowing what they are doing
● they are so nervous that most won't even reply!

Why don't you write – see if you do better than us. At least two retailers are launching pesticides initiatives.

Residues are usually present only in tiny amounts. Some people say that if it was not for superb laboratory equipment they would not be found. We say that, now equipment can measure the residues, it is time these cocktails were researched. Most research at present is carried out into individual pesticides, not combinations – next to no research is done into the cocktail effect. In reply to all these comments: the government says: '. . . occasional exposure to higher-than-average levels of pesticides in foodstuffs has no public health significance'.[27]

In the USA the National Toxics Campaign, an environmental group, runs its own pesticide residue testing scheme as part of its service to community, environment and worker groups. The group's Citizens' Environmental Laboratory can test for up to 56,000 different substances.[28] We would like to see such a service in the UK too.

Organochlorines will show up in both people and food, because they are persistent but less toxic. In contrast, organophosphates are less persistent but more toxic. We would like to see testing schemes for those who are most exposed to the persistent pesticides, such as wood preservatives.

When the National Health Service was first set up in 1948, there were plans to include a large Occupational Medical Service. This never happened. Until such a service is set up we would like to see trade unions and community groups in 'at risk' areas and industries campaigning for residue testing.

We say

- there should be research on the cocktail effects
- monitoring does not get rid of pesticide contamination
- there should be *no* contamination of food by pesticides
- information should be made public.

How Pesticides Affect Your Health

Symptoms

Have you any reason to suspect that you or your family or friends may be suffering from pesticide poisoning? Some of the symptoms have already been mentioned in earlier chapters: they are many and complex and range from headaches, dermatitis, muscle twitches and allergies through to reproductive damage, cancer and even death. The effects of the main pesticides are given in Table 2 at the end of this book, while the main symptoms associated with poisoning from the major groups of pesticides are listed in Table 4. If you have been exposed to a pesticide and you have any worries, consult your doctor. Don't leap to conclusions. Keep a record of your symptoms in a diary, and find out any information you can on what the pesticide was. See Further Reading for other books to consult. Take advice from your union, professional body or family doctor, and if you want to complain, read Chapter 14.

In the rest of this chapter we look at the effects of pesticides on health in both the short term and long term.

Short-term effects and the problem of under-reporting

Pesticides are designed to be toxic to living things. A recent review has put pesticide poisoning in perspective: 'The primary hazard of pesticide exposure is the development of acute toxic reactions as a result of dermal contact with, or inhalation of, a relatively large dose.'[1] Over-exposure by skin contact or breathing in pesticides can lead to acute effects ranging from eye and upper respiratory tract irritation and contact dermatitis to more serious poisoning.

Those at greatest risk are those who regularly work with pesticides, rather than people who use pesticides in the home and garden. Dr John Bonsall, chairman of the health and safety committee of the British Agrochemical Association, considers that 90% of operator contamination during spraying work occurs on the hands when opening containers and transferring concentrates.[2]

Under-reporting

In 1986 the National Poisons Unit suggested that up to five thousand people in England and Wales might suffer from acute pesticide poisoning.[3]

HSE now publishes an annual report called *Pesticides Incidents Investigated*. The report was first produced after prolonged representations from trade unions that many incidents were not being recorded. The trade unions also requested that a survey be carried out to try to estimate the difference between these reportings and actual incidents. They were told that such a study would be too expensive. Instead, HSE inspectors were asked for their own estimates of under-reporting.

The 1989–90 report notes: 'Many of the confirmed poisoning incidents were of people affected by drift from nearby spraying operations. It is seldom possible to qualify the extent of exposure in these cases, but it is of interest that low levels of contamination give rise to identifiable effects.'

The House of Commons Agriculture Select Committee, chaired by Sir Richard Body, considered in its report (known as the Body Report) that the under-reporting of incidents constituted a serious problem. Some acute symptoms are of such a frequent, mild and short-lived nature that they are difficult to ascribe to pesticides. Equally, Body continues, 'while it is true that many such mild symptoms may be wrongfully assigned to pesticide use, it is at least as likely that sickness caused by pesticide exposure may be ascribed to other causes'.[4]

The likelihood of under-reporting is increased since GPs are not necessarily trained to diagnose mild, or indeed acute, pesticide poisoning. Dr Murray, again in evidence to the Body inquiry, stated: 'As a student, I had no toxicological training . . . how can my colleagues . . . have skills for which they have never been trained?'[5] The BMA's 1990 report goes a long way to encourage greater pesticide awareness among doctors. The situations in which many poisoning incidents occur contribute to the difficulty of reporting them. Many occur in the home after timber treatment. Others occur as a result of exposure to spray-drift, the source of which may not be apparent.

HSE include in their report a map of the country, showing the locations of poisoning incidents. The Pesticide Exposure Group of Sufferers (PEGS) have also produced a map showing the locations of incidents of pesticide exposure reported to them. In any given area, PEGS notes on average ten times more incidents than HM Agriculture Inspectorate.

In 1989 a coalition of environmental, consumer, trade union and pesticide manufacturing bodies pressed for a National Poisons Incidents Monitoring Scheme, so that a comprehensive picture can be built up of the nature and incidence of pesticide poisonings. A welcome response from government has been to commission a feasibility study of how the present reporting systems can be combined and improved.

Long-term effects

Can they be assessed?

Any long-term effects of pesticides are extremely difficult to gauge. Unlike pharmaceuticals, pesticides cannot readily be tested on humans because of the ethical implications. This creates problems, because the effects in different species may well vary considerably. In giving evidence to the Body inquiry, Dr Murray of the National Poisons Unit expressed the view: 'After all, it has been said that one well-documented case of poisoning in a human is equivalent to 20,000 animal experiments.'[6]

One of the groups giving evidence to the Body inquiry confirmed how difficult it is to predict from animal testing how a chemical will affect humans: 'The implications for man in quantitive terms of virtually any long-term test done on animals are not known.'[7] Comparisons within and between animal species, and comparisons between animals and man, are difficult. 'Some studies have been undertaken, attempting to extrapolate data between different animal laboratory species, and between laboratory animal species and man. At present, there is no reliable method for predicting the actual type of toxicity that will develop in different species in response to the same chemical substance.'[8]

Other methods are now being developed to try to reduce the reliance on dosing batteries of rodents to death. Alternatives include methods involving bacteria, cultured animal cells, fertilised chicken eggs and frog embryos.

Increasing attention is also being paid to monitoring pesticides in use. Rats, however, have some difficulty in reporting headache, nausea or dry eyes. There is now an increased emphasis on occupational health, hygiene and the safety of pesticides in use.

Reproductive hazards

One of the first instances of the reproductive hazards of certain pesticides concerned the insecticide DBCP. It has been confirmed that it affects male fertility, but it continued to be exported from the USA and used in banana plantations in Costa Rica in the 1970s after this fact became known. (See Chapter 7 for more details). A number of other pesticides or groups of pesticides have been linked with causing reproductive hazards, but without any such clear outcome.

Behaviour

Pesticides may also have neurotoxic effects – effects on behaviour. The best-known group of pesticides with chronic as well as acute neurotoxic effects are the organophosphates, which were derived from the development of nerve gases in the Second World War. Pesticide-induced delayed neuropathy (deadening of the nerves) has been relatively well documented, while other behavioural effects caused by changes in the nervous system are coming to be recognised.

Concentrating on the neuropathological effects of pesticides and ignoring the behavioural symptoms of chronic exposure has led to the setting of occupational exposure standards at levels that could put workers at risk. Behavioural symptoms should have been observed as indicators of pesticide toxicity. One author argues that 'sensitive measures on a small number of subjects may provide data for risk assessment that are far superior to cruder data extracted from battalions of rodents'.[9]

Immune system

The immune system plays a major role in protecting the host from infectious diseases, and, arguably, from cancer.Within the same group of chemicals, some that cause cancer suppress the immune system, whereas other chemicals in that group that do not cause cancer have no apparent effect on the immune system. Exposure to carcinogenic pesticides, it has been argued, could potentially result in damage to the immune system.[10]

There are at least 14 pesticides, including members of the organochlorine, organophosphate, carbamate and organotin classes, and the chlorophenoxy herbicide 2,4-D and their breakdown products, which can lead to the alteration of normal immune functions.[11] An example has been reported from the USA involving Aldicarb, a carbamate pesticide sometimes found in US ground water. Women taking in low levels of Aldicarb-contaminated ground water over a period of time were found to have altered immune systems. Fortunately, no increased incidence of infection associated with the exposure was found, but longer-term studies were required.[12]

Can Pesticides Cause Cancer?

The quick answer is yes. But the big questions are: Which? How much? and To what? Carcinogenic or cancer-forming effects can be found in a number of ways.

Mutagenic tests find out whether genes are damaged by the chemical. There are several such tests, the most famous being the Ames test. This was developed by Professor Ames, who now says that natural chemicals cause more cancer than 'man-made' ones (see Chapter 14). Other tests look for chromosome damage. These tests are carried out according to World Health Organisation guidelines.

Long-term animal tests usually compare the effect of a pesticide at high and low feeding levels, or high and low inhalation levels. These levels are set to try to make the results relevant to possible intake for people. Long-term animal tests generally look for effects of a pesticide on growth rates, general bodily condition and any cell changes. Such tests are very expensive – in time, money and animals.

Teratogenic effects are those which may cause birth defects. Does the pesticide lead to unusual changes in the offspring of the animal exposed to the pesticide? These tests by their nature are long-term and even more expensive.

Animal studies have implicated:

• about 50 pesticides, including Benomyl, Captofol, Captan, Carbofuran, Lindane, Maneb, Thiram and Ziram, as causing cancer

- over 30, including Dinoseb, Diquat, MCPA, Paraquat and Propachlor, as causing birth defects
- Over 60, including Aldicarb, Atrazine, Chlorfenvinphos, Cyanizine, Dimethoate and Simazine,[13] as causing genetic defects
- Over 60 as having various reproductive effects, some of which are teratogenic (cause birth defects).[14]

What is the value of animal testing?

The general reassurance about pesticides given by industry and government is that they are safe because they are tested on animals first. What happens when adverse effects are reported from animal testing, such as the occurrence of cancers in rats or mice?

Predicting is difficult. There may be relevant differences between humans and animals. Particular strains of rodents may be susceptible to particular tumours in certain parts of their bodies, and may therefore 'over-predict' cancer risks. In discussing the toxicology of the EBDC pesticides, the UK Advisory Committee on Pesticides concluded that there were significant differences between the thyroid gland of the rat and the human thyroid. Tumours in the rat thyroid did not mean that humans were also at risk of thyroid tumours. The reverse has also been known to occur. Inorganic arsenical pesticides which have not been used in the UK for some years are not carcinogenic in animals, but are found to be so in humans.

What is the mechanism by which a cancer or tumour is caused? Not all chemical carcinogens appear to react with DNA, the genetic material of the cell. Those that do are termed genotoxic. Genotoxic chemicals may cause damage to cells and thereby cause cancers at doses that are less than toxic. Non-genotoxic chemicals may have 'threshold' doses – a level at which they become toxic, but below which they are not – and their effects may be reversed. There *may* be 'safe' levels of non-genotoxins. There may be no safe levels of genotoxins.

Animal testing is expensive and time-consuming, so high doses are used to establish a response. They may or may not establish a relationship between the dose of a chemical and the body's response.

Cancer and the 'drins'

The complexity of making judgments about pesticide safety is illustrated by the arguments about the cancer effects of two pesticides, Aldrin and Dieldrin, since the mid-1970s.

Tests done by the manufacturer on Aldrin and Dieldrin – known as the 'drins' – indicated tumours among mice. Shell, the Anglo-Dutch chemical giant, defended the drins in hearings at the Environmental Protection Agency (EPA) in the USA. High levels of Aldrin and Dieldrin had been fed to mice to make sure they picked up any effect. The EPA administrator, Russell Train, noted that tumours occurred at the lowest doses tested, so there was no safe level of exposure. The EPA decided that these substances should be suspended because of 'imminent hazard'.[15]

But the response in the UK to this historic decision was cautious, to say the least. No ban was proposed, as it was expected that the chemicals would be phased out because of their persistence – they stay around for too long. The UK authorities claimed that 'there is no direct evidence of Aldrin or Dieldrin causing cancer in man [*sic*] nor, with one exception, in animal experiments'.

According to the *British Medical Journal* at the time,

the evidence on which the suggestion of 'an unreasonable risk of cancer' is based is the production in mice given diets containing Aldrin and Dieldrin of tumours in the livers. . . . It is a matter of opinion how far the production of these liver tumours in mice is a reliable index of a hazard to man. It is not a matter of the credibility of the witnesses: the facts are not in dispute, it is their interpretation that is a matter of debate. The EPA's inquiry was held by a judge on its staff, who rejected the evidence of one expert witness who had served on committees on the safety of insecticides. In the absence of any new evidence that Aldrin and Dieldrin are carcinogenic to animals or man there seems no reason for the authorities in this country to follow the American recommendation. Furthermore, where there is no consensus of opinion on interpretation . . . the procedure adopted by the EPA for reaching a decision is not one we should attempt to imitate.[16]

This saga, which is continuing, underlines the need for more independent scientific research into pesticide safety.

Three views on the cancer risk

Few commentators seem to agree about what is or is not a carcinogenic pesticide. The lack of consensus covers academics, regulators and international agencies. There are several classifications of the carcinogenicity of pesticides, produced by different international bodies.

IARC

The International Agency for Research on Cancer (IARC) based in Lyons, France evaluates chemicals by means of its Working Group on the Evaluation of Carcinogenic Risks to Humans. After reviewing evidence, an overall evaluation is given and the pesticide placed in one of five categories:

- Group 1 (carcinogenic in humans)
- Group 2A (probably carcinogenic)
- Group 2B (possibly carcinogenic)
- Group 3 (not classifiable)
- Group 4 (probably not carcinogenic in humans)

The recent updating of the IARC's series of evaluations lists only two in Group 1 (arsenical pesticides, and vinlyl chloride).[17] It lists one probably carcinogenic pesticide (ethylene dibromide), 17 possibly carcinogenic pesticides and 26 not classifiable pesticides. Only one substance has gained the accolade of 'probably not carcinogenic'.

IARC also describes the way it assesses evidence for carcinogenicity. The evidence for carcinogenicity is labelled 'Sufficient', 'Inadequate', or 'Limited', or if there is no evidence of carcinogenicity it is categorised as 'Evidence suggesting lack of carcinogenicity'. The instances of Inadequate or Limited evidence are widespread throughout the classification. However, the most common category of evidence is simply 'No adequate data'.

IARC comments that it 'cannot establish that all agents that cause cancer in experimental animals also cause cancer in humans, nevertheless in the absence of adequate data on humans, it is biologically plausible and prudent to regard agents for which there is sufficient evidence of carcinogenicity in experimental animals as if they presented a carcinogenic risk to humans'.[18] IARC does not have the resources to review all chemicals or all pesticides. The pesticides reviewed by IARC form but a small fraction of those causing concern, and an even smaller fraction of the number of active ingredients in use in the UK. It is frequently unable to evaluate the most widely used or best-established pesticides due to data gaps – the information simply doesn't exist – and does not consider 'newer' active ingredients for many years.

A survey was conducted of those pesticides which were the most important economically in the world. For this sample of 72 pesticides of major worldwide importance, IARC was able to provide cancer classifications for only three. US EPA cancer classifications were available for 18 of those pesticides.[19]

EPA

The EPA based in Washington DC, USA has also made an attempt systematically to review the pesticides it suspects are carcinogenic. According to the EPA's classification there are no compounds classed as 'human carcinogens'. There are 30 'probable human carcinogens'. Of these, one case, cadmium compounds, is considered to show that there is 'sufficient evidence of carcinogenicity from animal studies with limited evidence from epidemiological evidence'. For 29 compounds there is 'sufficient evidence of carcinogenicity from animal studies with inadequate or no epidemiological data'. There are 35 compounds classed as 'possible human carcinogens' with 'limited evidence of carcinogenicity in the absence of human data'. Three compounds are 'not classifiable as to human carcinogenicity' due to 'inadequate or no human and animal data for carcinogenicity'.[20]

By April 1989, the US National Toxicology Programme and the National Cancer Institute had completed chronic toxicity/carcinogenicity studies on 63 pesticides. Twenty-six of these were considered to show varying degrees of carcinogenicity in animal studies. That was 41%.

Q Scores

In this book, you will come across some committees with amazing long titles. Here is the all-time winner, from the USA: the Committee on Scientific and Regulatory Issues Underlying Pesticide Use Patterns and Agricultural Innovation, of the Board of Agriculture of the National Research Council.

The committee was asked by the EPA to work out a mathematical model for assessing possible cancer risk. A pesticide's oncogenic potency is given a kind of score, expressed as a 'Q*'. The Q* is a statistical score, measured from a graph, and represents the change of tumour incidence as a result of change of dose in high-dose animal test. It does not matter what sort of tumour is found, or the site of the tumour on the body of the animal. A high Q* shows more tumours. A low Q* score indicates a weak response. In the USA, there is a debate about how useful and accurate Q* scores are. We believe, with the American Consumers' Union, that Q* scores are at least an attempt to set a consistent method for calculating risk.[21] There is no comparable programme of cancer studies in the UK. Nor is there any available list of data gaps or cancer classifications for pesticides in use. The inference is that British regulators rely on information from abroad first. Or else they review problem pesticides on a case-by-case basis.

For further information on the testing of pesticides see Chapter 9.

It's a Global Problem

'**U**p to 100,000 Thai farmers could have been poisoned by walking barefoot through rice paddies full of toxic residues of pesticides, a study said yesterday,' ran a Reuters' report in 1990.[1] Without knowing it, you may eat and breathe pesticide from far away. And the situation must be far worse for the workers themselves. Do you ever stop and ask what is going on?

A world in your mouth

Every time you eat a chocolate bar a whole world passes between your lips. Cocoa from one side of the world is mixed with sugar from the other side. Milk from nearer to home can be mixed with nuts and raisins from near and far. There may be tiny amounts of pesticide residues in all these ingredients. In 1990, a study by the *Mail on Sunday* newspaper tested four top-selling UK chocolate bars. It found very small amounts of Lindane residue – around a fiftieth of a milligram for every kilogram of chocolate.[2]

While these residues do tell a sober story to consumers, they indicate a threat to workers. In 1989, Dr Alastair Hay of Leeds University visited Brazil to see the conditions of the workers who spray pesticides on the cocoa plantations. One region had had 300 pesticide poisonings and 14 deaths in 1986–8. Dr Hay said: 'This is just the tip of the iceberg. . . . These figures are a gross underestimate of the problem.'[3]

The Dutch Food Workers' Union has set up the International Cocoa Workers' Programme, linking various groups including the International Union of Foodworkers and the International Federation of Plantation, Agricultural and Allied Workers with the aim of pooling their experiences. Reports from Malaysia and Brazil of heavy chemical use, lack of medical care, repression of trade unions, no training, and labels only in English led to questions being put to the big multinational companies. Do they test for residues, and if so what are the results? International organisation also prevented Brazil's cocoa authority, CEPLAC, from sacking or transferring the main organisers of the cocoa technicians' trade union.

UK chocolate companies started testing for pesticide residues in 1972.[4] Residues in British-made chocolate are well within current safety limits, for the vast majority is made from West African cocoa which the Biscuit, Cake, Chocolate and Confectionery Alliance (BCCCA) considers is produced more safely than in Brazil. The BCCCA also says it backs best practice. But the insecticide Lindane is widely used in Ghana, the main West African cocoa-producing country, despite bans in certain other countries (see Table 2). In the Netherlands, the Cocoa Workers' Programme supports pesticide-free cocoa products. Pesticide-free coffee from Mexican small farmers is widely available in Dutch supermarkets.

Warning! Pesticides in transit

Pesticides are always being moved – around the globe, between suppliers and farms, in trucks and on ships. So it is not surprising that there are constant incidents involving pesticides in transit. Here are just a few of those reported in a survey conducted by the United Nations in 1967. As far as we know there has been no more recent survey of this kind.

- In the Middle East during 1967 there were two separate incidents of Endrin leakage on to sacks of flour in the holds of ships. They resulted in a total of 26 deaths, and more than 800 people who ate bread made from contaminated flour became ill[5]
- In Colombia in 1967 63 people died and 165 became ill from eating food made with flour contaminated with Parathion during transportation by lorry[6]
- In Texas, six members of one family became seriously ill when they ate tortillas made from flour contaminated with the organophosphate compound Carbophenothion.[7]

The survey concluded that there was serious under-reporting of pesticide safety problems related to transportation and storage: 'from conversations with warehousemen and drivers and from personal experience with spillage incidents . . . there is a tendency to try to 'clean up' any spillage without contacting the shipper or pesticide company or obtaining other expert advice. The result is inefficient decontamination. . . ."[8] One case suggests that matters may not have changed much. Eleven people from Dominica who stowed away in the cargo hold of a ship were killed by a pesticide sprayed on the cargo of cocoa in Puerto Rico.[9]

Ecological dumping

Pesticides are a good illustration of how the world is one entity. European-based companies sell pesticides abroad. Food is grown using them, sometimes with few or inadequate controls. Companies say they do their best: 'It's not our fault if they don't read the labels.' (But are the labels always in the language of the country where the chemicals are to be used? No.) Then the Ministry of Agriculture, Fisheries and Food surveys pesticide residues and says that it is concerned mainly about residues in imported food.[10]

Increasingly, UK pesticide companies are looking to export markets to increase sales of their products. In Chapter 8 we give you the figures on sales of pesticides. But are Third World countries given advance notice that a pesticide may have safety problems? The worldwide Pesticides Action Network (PAN) successfully argued in favour of what is called Prior Informed Consent (PIC).[11] PIC has been adopted into international pesticide convention (see Chapter 11). It means that a company exporting a pesticide would have to release all its safety data. This would stop 'ecological dumping' where. rich countries try to clean up their environment by dumping their unwanted waste and outdated products on less well-protected countries.[12]

Mauritius, a small country of just over 1 million people, has the highest pesticide consumption by weight of active ingredient (i.e. pesticide before dilution, excluding a carrier such as water) per hectare in the world. Its agriculture is dominated not by food for home use but by sugar cane for export.[13] In Colombia, death from pesticide poisoning has doubled in the period 1978-86.[14] In Paraguay, ICI advertisements for Gramoxone (Paraquat), gave no warning phrases or symbols or instructions to read the label. Yet it claimed that the farmer would have 'up to 23.6% more net profit per hectare'.[15]

The Pesticides Trust report says that of 25 'potentially very hazardous pesticides' some require special equipment and training for users. Despite this, they are being heavily traded in the tropics by both US and UK companies.[16] Controls need to be improved globally. 'Think global – act local' is the Green slogan. Perhaps we could add: 'Co-ordinate international'!

International pressure to reduce use

All over the world, people are reacting to this global challenge in similar ways: monitoring, educating, asking, sharing information. Consumer, trade union, environmental and public health organisations are getting together and asking questions. Why weren't we told about residues? Who says the risk is low or high? What about our children? What do the workers have to put up with?

Fruit on sale in UK supermarkets can be the pick of the world. Before Daminozide was withdrawn in 1989, Parents for Safe Food commissioned tests on 26 apple samples or apple products. One Red Premier apple from New Zealand, where since 1985 the Apple and Pear Marketing Board has instructed growers not to use Daminozide, showed up Daminozide residues (4 milligrams per kilogram). Discussions were held with New Zealand fruit marketers to find out what had happened. The biggest residue of all was reported for a Bramley apple. Daminozide is now being phased out after consumer pressure worldwide. It is still legal in the UK.

In Queensland, Australia, a fruit growers' marketing company responded to global consumer pressure about pesticides. In 1990 it started marketing not only Daminozide-free apples, but also apples and pears without wax spray coatings containing fungicide. In addition, the Australian Consumers' Association set up a Working Party on Pesticides, whose members comprised scientists, environmentalists, representatives of public health interests and consumers. They have been negotiating a complicated deal with farmers, growers and retailers to reduce pesticide use.[17]

Pesticide reduction policies have been introduced in Sweden, the Netherlands and Denmark.[18] If these countries can aim to cut use by 50% in a few years, why can't the UK? Pressure is mounting. Have you ever thought about bananas? People in

Japan did not give them much thought – until a Filipino banana worker went to Japan to talk with consumer groups in 1980. Public attention focussed on the pesticide residues. Others were also concerned about the liberal doses of chemicals dished out to the plantation workers. The Stop the Philippine Banana Pesticides Campaign, set up in 1985, found measurable levels of Benomyl, thiophanate-methyl, TBZ and Mancozeb in the bananas. Some of these residues were not picked up by the Japanese authorities because inspection were either left to each importer or carried out for only 26 of the 300 pesticides approved in Japan.[19]

A leader of the Filipino banana workers' union visited Japan to tell of the plight of chemical applicators on the plantations. Campaign members visited them and found workers handling pesticides without gloves or masks. They complained of wrinkled skin, stomach aches, poor appetite and even sterility. The investigators found 26 kinds of chemicals, including Aldicarb, Phenamiphos, Ethoprop, Paraquat, Fenitriothion and DBCP.

DBCP was never approved for use in the UK. Animal tests carried out in the 1950s showed that DBCP could destroy sperm-producing tissue. It is a World Health Organisation class Ia pesticide (see p. 183). In the late 1970s some men involved in manufacturing DBCP at the Occidental plant in California were concerned that, although they wanted children, none of their wives was becoming pregnant. They called in their union, the Oil and Chemical Workers, who persuaded all the workers to undergo medical checks. Fourteen out of 27 were either sterile or had very low sperm counts. Workers at Dow Chemicals in Arkansas were then checked: 62 out of 86 men were found to be suffering from the same condition. Other DBCP plants produced the same results. Other tests found that DBCP caused cancer in test animals.

The US authorities then imposed lower limits of exposure and set 'food monitoring' to find out the contamination of food already on the market. The *New York Times* was moved to say: 'Must we continue to make belated discoveries? It was the workers, not the regulators or manufacturers, who prepared the indictment.'[19] Robert Phillips, executive secretary of the National Peach Council, put another angle on it. 'There can be good and bad sides to a situation. Could workers be advised so some might volunteer for such work posts as a means of getting around religious bans on birth controls?'[20]

Imagine how the Filipino workers felt. They had to endure a danger that had already been discovered in the USA ten years previously. But pesticides travel faster than news of their hazards. Since then the Filipino Campaign has collected further evidence and now makes safety part of the workers' negotiations. It emphasises the reduction of chemical use and argues that banana consumption would increase if consumers learned that the bananas were free from chemicals. That in turn would increase profits. A separate organisation now produces organic bananas for export, and by 1989 60 tons of organic bananas were being exported to Japan every year.[21]

PAN – the Pesticides Action Network

There are many other tales of international connections and cooperation, and in the 1980s there was an increasing awareness to coordinate these links. PAN is an international coalition of groups and individuals who oppose unnecessary use and misuse of pesticides. It supports reliance on safe, sustainable pest control methods. PAN links 300 organisations in 50 countries, via seven regional centres.

In 1985 PAN launched the Dirty Dozen campaign. It selected a number of pesticides for strict controls, bans and ultimately worldwide elimination. Groups from 60 countries picked the pesticides according to human hazards, environmental impact, usage, and bans in exporting countries. The campaign's goals are:

- to ensure human safety and environmental health
- to end the use of the Dirty Dozen wherever safe use cannot be assured
- to eliminate double standards in the global pesticide trade
- to generate research and support for sustainable pest control methods.

The Dirty Dozen actually comprise 13 pesticides or groups of pesticides (Aldicarb was added later, making a baker's dozen!). They are:

- Aldicarb
- Camphechlor (Toxaphene)
- Chlordane and Heptachlor
- Chlordimeform

- DBCP
- DDT
- the 'drins' – Aldrin, Endrin and Dieldrin
- EDB
- HCH and Lindane
- Paraquat
- Parathion and Methyl Parathion
- Pentachlorophenol
- 245 T.[22]

By working together, groups are more likely to bring about a change to more just and sustainable production systems. (See also Chapter 11.)

The Vast Pesticides Market

You might think it doesn't matter who makes a pesticide. We think it does. Companies differ in their morals, size and power. Chapter 2 tells you more about their history; here we want you to know who the companies are, and how big they are. If you buy pesticides or eat their residues you are dealing with a tiny number of companies, but theirs is a worldwide market.

A concentration of power

Immediately after the Second World War, IG Farben, the German chemicals giant, was broken up by the Allies. Three companies were formed – Hoechst, Bayer and BASF – and each is now in the top ten of world manufacturers in its own right.

During the postwar period the industry grew and prospered. Between 1945 and 1950, 20 companies entered the market and 28 new pesticides were produced. By 1960 50 companies had joined the field, as had 145 compounds. By 1970 500 pesticides had been introduced and 75 companies were screening chemicals for pesticidal activity. Some were petrochemical companies using their chemists to find new pesticides. Others were pharmaceutical companies making all sorts of compounds.[1]

Here is the ranking of the top 20 world agrochemical companies. The list is based on sales, mainly pesticides. The figures are for 1989.[2]

Company	Country	Domestic currency (million)	US$ (million)
1. Ciba-Geigy	Switzerland	SwFr 3,713	2,271
2. ICI	UK	£ 1,238	2,026
3. Bayer	Germany	DM 3,500	1,862
4. Du Pont	USA		1,684
5. Rhône Poulenc	France	Fr 10,500	1,646
6. Monsanto	USA		1,558
7. Dow Elanco	USA		1,485
8. Hoechst	Germany	DM 2,050 (3)	1,090
9. BASF	Germany	DM 1,940	1,032
10. Shell	UK	£ 552 (3)	903
11. American Cyanamid	USA		820
12. Schering	Germany	DM 1,392	740
13. Sandoz	Switzerland	SwFr 1,165	713
14. Kumiai	Japan	Yen 53,858	420
15. FMC	USA		414
16. Rohm and Haas	USA		367
17. Sankyo	Japan	Yen 47,500	344
18. Nihon Hohyaku	Japan	Yen 43,854	318
19. Takeda	Japan	Yen 37,000	268
20. Hokko	Japan	Yen 36,511	265

The top 10 agrochemical companies accounted for around 72% of the world market in 1989. The top 20 companies held 94% of the world market.

The world agrochemical industry is concentrated in the industrialised nations, and only a few countries are represented: the USA and European countries dominate the top ten. It seems unlikely that any more European companies can enter the big league, although the industry is well represented in Scandinavia. The most likely additions will come from Japan.

Research and development

With the popularity of pesticides on the rise in the 1950s, 1960s and 1970s, companies were constantly screening for new chemicals. Some were required to combat pests that had become resistant to the earlier chemicals. A new chemical was the manufacturers' way to solve a problem created by an old one. In the late 1960s 25 pesticides were introduced each year.

By the late 1970s new additions had fallen to 16. It took three times as many chemicals to find one useful one, and the costs of development had risen dramatically.

Here's how the industry's outlay on pesticides broke down:

- research and development 7%
- production costs about 50%
- fixed capital costs about 5%
- technical service about 4%
- administration 15%

That left

- profit about 20%.[3]

Making pesticides on contract

The production of pesticides involves two main stages. The active ingredients are manufactured first, often by smaller companies contracted to the major named companies. This is because the many types of pesticides require relatively small production lines or 'batch' processes.

The second stage is formulation – the processing of the basic chemicals into a usable form such as spray, powder or mixture of chemicals. Formulation involves more small companies, although they are sometimes one and the same.

As an example, Bendiocarb is marketed in the MAFF-approved list (see p. 214) by three companies – Camco Ltd, FBC Ltd and Schering AG. All belong to Schering. The large companies are moving in to take over more of the formulating companies. There are about 25 UK formulation companies, as distinct from the major manufacturers.[4]

All change in the 1990s?

In the 1990s the industry faces a real problem – declining rate of profit. Higher oil costs, fewer chemicals and the need to find new ones all mean companies have to raise prices. This, plus increased pressure for tougher environmental laws worldwide, has made the pesticide companies invest elsewhere – in exports and in buying each other out. We think that in the 1990s pesticide companies will have a three-pronged strategy: the 3 Bs –

buy-outs, bale-outs and biotechnology.

The companies' attention is now increasingly focussed on seed companies and biotechnology. They want farmers to use plants with built-in resistance or varieties that can only flourish with the use of specific pesticides manufactured by that company. The first UK approval for release of a herbicide-resistant sugar beet seed was made in 1990. Monsanto and Maribo, the companies involved, say this will be more economic and environment-friendly. We are not so sure.[5]

Bale-outs happen when a company decides its profit margins are being squeezed and it is better to sell off a division or subsidiary. The gradual concentration of world market share is evidence of this happening.

The third option is buy-outs and mergers, but they too can come unstuck. In 1990 Sandoz and Schering planned a joint agrochemical venture which would have put them in the production top 10. Sandoz would get better access to European markets, and Schering better access to US markets. Together they expected to penetrate eastern Europe. Suddenly it was all off. Circumstances changed, and the plans were put back into the filing cabinet.

Today's market

The market for pesticides is huge both in Britain and worldwide. When you reach for a container of slug pellets on the shelf of your local garden centre, you too are joining that vast market. The total world market in agrochemicals in 1989 was worth around $21,500 million.[6] Here are some more mindboggling figures for you to digest.

Between 1945 and 1960, worldwide sales grew 7% each year. Sales between 1962 and 1970 jumped 15% a year due to higher prices. In the 25 years up to 1983, the average increase was 12.5% per year.[7] And so it goes on. There was a 3.2% increase in world demand in 1989. The reason? The area of land planted with food crops and cotton increased by 1.4%. The USA saw a significant increase in planted acreage – up by over 10% for wheat and maize. To balance this, the Japanese market decreased slightly and there was little growth in Latin America. But with starvation facing parts of eastern Europe and the collapse of the communist farming system, agrochemical companies expect that this region will provide a major growth market over the next few years.

In the UK, sales increased by 6.1% in 1989 over the previous year. Since 1983 sales have increased by 31.3%, which is a slight fall when inflation is taken into account. In the five years to 1989:

- herbicide sales increased by over 16% to over £200 million
- insecticide sales increased nearly 30% to £47 million
- fungicide sales increased by 47% to £135 million.

During these years total UK sales increased by over 25% to £433.1 million. Sales of pesticides going abroad during the same period increased by 37.2% to £655.5 million.[8]

Which pesticides are most commonly used in the UK?

One of the most difficult facts to discover about pesticides is which are the ones used most. If you think about it, such secrecy is stupid. The Society of Motor Manufacturers and Traders releases sales figures for the top 20 selling cars each year, and the top 10 of each month. Yet similar figures for pesticides are top secret. The British Agrochemicals Association doesn't keep such figures. An 'off-the-top-of-the-head' assessment given to us by one industry spokesperson went as follows. The most commonly used pesticides would be herbicides. And most of those would be autumn-applied. And probably around 100 products would account for 80% of sales in the UK.

Here are some industry figures we managed to unearth for you. The three most commonly used pesticides in each group of agrochemicals in the UK in 1987 were:[9]

Group	Active ingredient (common name)	Trade name (main ones)	Company (main ones)
Herbicides	Isoproturon	Arelon	Hoechst
		Sabre	Schering
		Tolkan	Rhone-Poulenc
	Metsulfuron-Methyl	Ally	Dupont
	Triallate	Avadex BW	Monsanto
Fungicides	Propiconazole	Tilt	CIBA Geigy
		Radar	ICI
	Prochloraz	Sportak	Schering

	Fenpropimorph	Corbel	BASF
Insecticides	Dimethoate	Many companies' names	Dimethoate
	Deltamethrin	Decis	Hoechst
	Pirimicarb	Aphox	ICI
		Pirimicarb	Schering

According to the Ministry's Agricultural Development Advisory Service (ADAS) the most commonly used pesticides in 1989 were:

- *herbicides:* Mecoprop, Isoproturon, Ioxynil, Bromoxynil
- *fungicides*: Carbendazim, Propiconazole, Prochloraz, Fenpropimorph, Tridemorph
- *insecticides:* Dimethoate, Demeton-s-methyl, gamma HCH (Lindane), Pirimicarb, Cypermethrin, Deltamethrin[10]

Are sales falling or rising?

The pesticide market is big business. It is ironical that these days agrochemical companies try to reassure critics by saying that sales of pesticide active ingredients (the concentrated form, before dilution) have dropped.

We accept that sales have fallen. All the figures agree. The big question is: by how much? The British Agrochemical Association, the industry umbrella group, says that the reduction is from 33,000 tonnes in 1983 to 24,000 tonnes in 1989. Yet the Ministry of Agriculture reckons the figure is 29,000 tonnes.[11] In a letter to one of your authors, the figure was given with amazing precision: 29,361 tonnes. No more, no less.[12]

Should you and we be happy to see this fall? Not completely. The falling sales argument is a bit of a smokescreen. Active ingredients may have dropped, but their power has gone up. Modern pesticides pack more of a punch – they are more 'specific', to use the industry jargon. Just as importantly, the area of UK land treated with pesticides has gone up dramatically. During the same period, 1983–9, that area has increased by over 50% – from 15 million hectares to 23 million hectares. This growth indicates that less pesticide is being used on more land.

This could mean that pesticides are getting more 'specific' or targeted and powerful. Or it could mean that pesticides are not winning the war on pests or weeds – that there is a 'treadmill' effect. The more that gets applied, the more dependent on pesticides the agricultural industry becomes, and the less effective these controls are. Look at what the pesticide companies themselves are finding.

Are the weeds winning?

Go to your local library and open a farming paper or magazine. If you look at the agrochemical company advertisements, you will find they promise great success, an end to weeds, bugs and problems. Just how effective are they?

Schering, one of Europe's pesticide giants, revealed in 1989 that a weed survey of 2,500 fields threw up the following facts:[13]

- the top four weeds – chickweed, speedwell, mayweed and cleavers – were showing a remarkable resilience to pesticides
- some weeds had gained ground since the last big survey, in 1967
- in 1967 chickweed was found on 77% of fields. In 1989, chickweed was reported on 94% of the fields surveyed.

Another agrochemical company, Ciba-Geigy, found from a survey carried out in 15 cereal-growing counties of Britain that:[14]

- cleavers and field pansies were the main problems for 40% of farmers questioned
- 27% of growers thought blackgrass, a major weed problem in cereals, was on the increase, 22% felt herbicides had it under control, and 10% were dissatisfied with herbicides for blackgrass.

The government's Agricultural Development Advisory Service has also found:[15]

- one in ten samples showing some sign of resistance to herbicides
- none of 14 herbicides it tested giving more than 75% control
- £4.1 million-worth of spraying was a complete waste of time.

Herbicide-resistant weeds have now been identified in all but ten states of the USA, in eight of Canada's ten provinces, in 18 European countries and in ten other countries too. The number of herbicide-resistant weeds has tripled since 1982, and the amount of infected land gone up by ten times.[16]

This particular story deals with herbicides. The same story was told for insecticides ten years ago. The result is not surprising. *Farmers' Weekly* reported a survey of 1,660 farmers and landowners which found that over half believed their colleagues were using too many pesticides.[17]

In a survey of 12,000 EC consumers in 1987, 81% agreed that the use of fertilisers, weedkillers and pesticides should be reduced – even if that meant paying more for the produce.[18] We agree that too much pesticide is being used. Why? Because pesticides are not as beneficial as their propagandists say. Because pesticides are working less and less well. Because they are not wanted or needed as they once were. And because they pose unnecessary risks, about which users and public alike are not consulted.

9
How Are the Dangers Judged?

'**S**ure, Chlordane's going to kill a lot of people, but they may be dying of something else anyway.' This statement was made by a US pesticide regulator.[1] The assessment of dangers (touched on in Chapter 6, specifically in relation to cancer) is at the heart of all regulatory action. Risk assessment involves all manner of technical, ethical and social issues. The nature of the danger has to be established. Then there is the measurement of risk. Having worked that out, the type and likely extent of usage have to be determined. Complex decisions have to be made with regard to the current climate of expectations. Nobody says risk assessment is an easy process. But here is a guide to what goes on.

The UK Advisory Committee on Pesticides (ACP) explains in some detail how it assesses toxicity. The Toxicological Assessment from its Report of 1988 gives in ACP's own words a complete rundown of its method, and is obtainable from HMSO. Few other authorities have spelt out their methods as clearly. Thus our comments are aimed not specifically at the ACP but at the general principles behind all such assessments.

The ACP Toxicological Assessment sounds good and looks objective, trying to build fact upon fact in a constant search for truth. Its integrity depends upon the scientific method. Responsible and respectable people work on your behalf to protect your safety. Standards are determined by groups of experts who make decisions after extensive studies. The data on which to base those decisions may not be all you could hope for, but the experts representing the relevant academic disciplines know how to interpret it. The message reassures you that you can trust them.

Their message, however, is difficult to check. When you do try to check, you find there are two basic issues that need further thought. How good is the data on which risks are assessed? And how good are the decisions?

Collecting human data

You have seen in Chapter 6 how the reporting of pesticide poisoning is not an accurate way of counting such incidents. One concern is that there is under-reporting – that many people who may think they are affected do not report the matter to the appropriate authority. So incidents do not reach the record book. Yet it can only be from the record book that you can find any records – the data which all those experts need in order to formulate measures to safeguard you.

Few surveys have been carried out on people who may be exposed to dangers. The most famous is one in which Swedish railway workers were surveyed and found to have an increase of soft-cell cancers[2]. It was considered that exposure to a mixture of pesticides may have been the cause. It was impossible to isolate which of several possible pesticides might have been responsible.

It is also hard to isolate a group of people who can be guaranteed not to have been exposed to any pesticide. So in formal scientific terms, there is no 'control' group which can be used to make comparisons between people exposed to pesticides and those who are not. Like the ACP, we recognise that it is virtually impossible to categorise people according to minute exposures to pesticides. Nevertheless we would argue that many more surveys should be done of people most likely to be affected.

We also accept that it would be completely unethical to experiment by spraying people to see what happens – although Ciba-Geigy did just that on one occasion. In 1976 the pesticide Chlordimeform was sprayed on unprotected Egyptian children to test its safety. Six children aged 10 to 18 were paid about $10 each to stand in a field, unprotected, while a plane sprayed them from a distance of five metres. A Third World lobby group discovered documents which examined how much of the pesticide was retained in the children's urine. The pesticide was subsequently linked with cancer. Ciba-Geigy conceded in December 1981 that it deeply regretted having used the children, Chlordimeform was withdrawn from sale in 1976 put back in 1978, and stopped by Ciba Geigy in 1987. It is one of PAN's Dirty Dozen pesticides.[3]

Then there are the risks from using pesticides in houses. A widely used pesticide called Fenitrothion is sold in the UK under about 20 trade names. The following summary of Fenitrothion's safety is edited from a report of the Expert Committee on Pesticides of the World Health Organisation. Look at how it made its judgment that Fenitrothion was safe to use indoors.

Studies were carried out during four rounds of Fenitrothion spraying undertaken. . . . Three different formulations were tested, each applied indoors. . . . A routine twice-weekly cholinesterase [see Table 4 under 'organophosphates', and Glossary] check was carried out. The spraymen's protective clothing consisted of overalls, broad-brimmed hats, tennis shoes, and gauze surgical masks. . . .
Slight to moderate depression of cholinesterase was determined in most spraymen towards the end of the spraying rounds. One sprayman was removed from the spraying schedule in the first round. Another was removed in both the second and third rounds of spraying. In the fourth round inhibition of cholinesterase was less pronounced; attributed to stricter safety precautions and improved supervision of workers. Except for one case of headache and slight dizziness due to 50% inhibition of cholinesterase on the final day no other complaints attributable to exposure were recorded during the 200 man-days.
No complaints whatsoever were received from the inhabitants, numbering 20,000.
A number of chickens were poisoned by drinking water contaminated with Fenitrothion. Arrangements for adequate disposal should be made.
On the basis of these results the Committee concluded that Fenitrothion could be used safely as a residual spray in houses, provided that precautionary measures similar to those described above are followed. . . . Cholinesterase determination should be carried out once a week whenever the spraying lasts more than four weeks. The Committee therefore supports the use of Fenitrothion.'[4]

Is that what you would call safe? Or a test for safety? We hope the actual decision wasn't as casual as the write-up implies.

Collecting animal data

All this puts great emphasis on to animal tests. Defenders of animal studies say they have to be good, and only done to the best standards. We agree. But how do we know?
The raw data usually comes from the manufacturing companies themselves. Little data comes from independent or government sources. The process that ultimately controls company use relies on their own data. Much then depends on that data.

Some companies have been less than forthcoming with their information. Velsicol, the company that makes Heptachlor and Chlordane, failed to submit critical unpublished data to the US Environmental Protection Agency. Such data included results of reproductive and teratogenic tests. The company also failed to pass on to the authorities the 1200 complaints it received annually, or details of blood disorders. Nevertheless it continued to say that these two pesticides posed 'no identifiable health hazard' and that Chlordane has an unmatched safety and performance record. Two directors were taped saying that if the results of tumour tests 'ever get to the public hearing. . . we wouldn't have a chance in hell'. Six company directors were indicted but the case was withdrawn. Some years later the EPA and Velsicol came to an agreement that Chlordane stocks could be sold off, and in return Chlordane's registration would be struck off the EPA register.[5] This shocking case may be the odd bad example. But trouble is that we cannot judge how unusual it is, because such information is commercially confidential.

Let's now take a look at some of the information you will encounter when assessing the dangers of pesticides. The full data requirements that the Ministry demands for any new pesticide are now huge.[6]

The LD50 test

This test measures the amount of a substance required to kill 50% of the animals used. The LD50 is a very crude measure that provides very little useful information, and has killed thousands of animals in its time. We wonder why it is the only freely available piece of information on all pesticides.

From tests on rats it can be seen that Chlormephos is more toxic than Parathion, which is more toxic than arsenic compounds (Paris Green), which are more toxic than strychnine, which is more toxic than nicotine and Dinoseb (recently withdrawn).[7] They are all more acutely toxic than Paraquat. LD50 tests give you no more than a very rough pecking order of danger, and that is why many scientists don't like it as a test any more.

How solid is the data?

In 1983 one of the major testing organisations, Industrial Bio-Test (IBT) of Chicago, was convicted for fraudulent practices –

including the fabrication of data. This meant that all its test results have to be formally judged as meaningless. The following are those pesticides on which approval rested wholly or in part on the IBT testing:

- Triallate – the third best-selling herbicide in the UK
- Bifenox
- Captan
- Carbofuran
- Dicamba
- Glyphosate
- Alachlor
- Picloram
- Propachlor
- Endosulphan
- Barban
- Chlorbromuron
- Terbutyn.[8]

All are on the UK approved list! The ACP writes about the need for the data for pesticides to comply with 'current scientific standards and acceptable testing protocols'.

How good are laboratory procedures?

No one fully knows how the IBT scandal altered the risk assessment of pesticide compounds in the UK or the USA. In the USA the Environmental Protection Agency stated that the data on which it would rely in future would be 'of impeccable quality'. The EPA promoted 'Good Laboratory Practices' that will enhance the quality of laboratory test data required for evaluation of chemicals. We do not know how other authorities ensure data of 'impeccable quality'.

What about the gaps in pesticide data?

The USA's National Academy of Sciences has estimated that the US authorities hold adequate safety data for only 10% of pesticides, incomplete data for 52%, and no information at all for 38%.[9] The regulations establish standards and a vigorous audit programme to ensure compliance. Whenever inadequate data is found, the Environmental Protection Agency will 'take an aggressive posture' towards correcting, will prevent further expansion of the use of the pesticide(s) in question and will demand prompt replacement of the missing data.[10] The UK review we hope will perform the same function in time.

Decisions

With something as complex as pesticide safety data, there are bound to be difficulties in agreeing what the data means. Bearing in mind these limitations, you can look at decisions and how they are made. There are a number of consistent features, listed below.

No single set procedure

The way standards for residues in food are set is described in Chapter 10. Whatever its faults, it has some internal consistency. That cannot be said of how standards for pesticide exposure in the workplace are met. Workplace levels were decided over 20 years ago by the American Conference of Government and Industrial Hygienists. Many levels were based on a 'guestimate' from a few studies on volunteers (see p. 87 for how the level for Lindane was arrived at).

Different authorities arrive at different conclusions

There are so many examples of the authorities disagreeing that we cannot begin to list them. Look at Tables 3 and 4 to see how different authorities come to different conclusions for the same pesticide, whether in food, in water or at work. There is much argument and debate within and between authorities as to:

- the *status* of any given set of tests. Attempts are made to standardise the tests, but they are continually thwarted. New sets of tests are continually being designed for particular sets of circumstances or pesticides
- the *relevance* of each test. Can the results of animal tests be adapted for application to humans? Not all chemicals that cause cancer in animals cause cancer in humans, and vice versa. Some authorities believe Lindane has a worse effect on humans than on animals
- the *confidence* given to one set of tests versus another. Numbers of animals, length of the test, and who is doing them all contribute to different assessments of how much each authority relies on the result

Possibly the greatest area of difference betweel
carcinogenicity (see Chapter 6). From Table 4 you
several major authorities – all hard-working, object.
entific – cannot agree on which pesticides are possibl\
cancer in humans. In 1989 many US farmers were tol   ɔp
spraying a number of EDBC fungicides linked with cancer.
They have been used in the UK for 40 years. The UK Ministry
admitted it knew of the US ban, but disagrees with the US
view.[11]

The criterion for decision-making is constantly changing

To quote the US EPA: 'The agency must clearly articulate the
process by which pesticide safety is reviewed and regulatory
decisions made. . . . During the last few years the process has
been in a state of flux and the uncertainty in the regulatory
approach has led to confusion.' (For more confusion in action
see Chapter 12.[12] New criteria, new demands for studies and
new definitions appear all the time. There is little consistency.)

Older pesticides are different from new ones

Within the same authority at the same time, pesticides can
receive different treatment. Very often new pesticides have to
go through tighter scrutiny than pesticides that have been in the
environment for ages. Their use may or may not be reviewed.
Or the review process may be conducted at such a slow pace
that the pesticides in question will have stopped making a profit
by the time their turn comes round. In 1989 a group of concerned UK voluntary bodies – the Pesticides Trust, the National
Federation of Women's Institutes, Friends of the Earth, the
Green Alliance and the Transport and General Workers' Union
– joined forces with the British Agrochemical Association to call
on the UK government to speed up its review of older pesticides. There has been little improvement. Old pesticides remain
in use, their safety based on outdated procedures.

In the USA nearly 70 out of 300 pesticides approved before
tougher procedures started in 1972 (following the 1972 Federal
Pesticide Act) have been found by the government's Environmental Protection Agency to cause cancer in animals. Ironi-
cally, according to Consumer Research Inc., the tougher regulations have allowed more dangerous chemicals to remain in
use while softer pesticides have been kept out of use.[13]

Prevention, not cure

Much of the debate about pesticide safety is a reaction to events. Few authorities go out and try to *predict* problems. The US EPA now claims to do that. It says: 'Rather than solely reacting to instances of contamination, EPA is identifying those pesticides that pose the greatest potential for contaminating water so that action can be taken *before* any problem occurs.'[14] Other authorities appear afraid to do this. They are either fearful of what they may find or fearful that such an approach will be seen as an admission that something may be wrong. Have you ever heard the statement: 'There is no evidence to show that pesticides cause damage'? Remember that the absence of evidence is not the same as evidence of the absence of effects!

The process of assessing pesticides was built up in an *ad hoc* way, just like the science and discovery of pesticides. Attention was focussed first on one pesticide, then on another. If there was a complaint, an inquiry would look into one particular aspect. It may have been birds' eggs one year, operator safety the next, water pollution after that. The problems posed by pesticides were not examined as an entity. If one was considered too dangerous for a particular use, this opinion would not be taken as an indictment of other uses. Only now is there a growing realisation that we need a consistent approach in order to assess pesticide safety.

The problem about risk assessment – currently in vogue with pesticide companies – is how to let consumers and users know the risks. Risks may indeed be small. Here are two examples – one produced by a state body and the other by a public interest group.

The US Food and Drug Administration tested fruit, vegetables and diary products for pesticide residues. It found that just under one in twenty samples had 'technical' violations of regulations. One in a hundred had levels just above the legal limits.[15] By and large the FDA was reported fairly happy with its findings.

In New Zealand, the Consumers' Institute found that many of the cereal samples it tested for pesticide residues exceeded the limits set by the government Food Regulations. It was not happy when reassured by health officials that this posed no danger to consumers.[16]

The point is: who says what is a danger for whom? How does a label get attached to that? What if you get the one in a hundred over the limit? What if you are the farmer using a pesticide which has not been tested by modern methods? (Farmers are caught in what psychologists call a 'double bind' here; they cannot win.) What if you or your baby are the one in a hundred thousand? The shift from being a tiny risk to being a victim is huge.

It is easy for anyone not at risk, or who escapes risk, to make decisions about risk to someone else. This may pass as objective science, but is it ethical? That is why at-risk groups must have some representation in open decision-making.

Time for a new approach

We argue that no single correct assessment of the dangers of pesticides exists. We do not believe it is simply a matter of some figures changing. A new process needs to be developed, and it should include:

- criteria that are clear and consistent
- the representation of public interests – workers, consumers, public health and the environment
- all information to be freely available without qualification
- structures which enable differences of perspective and information to be discussed openly and, it is hoped, resolved.

One possibility, the consensus conference, has been tried in Denmark. A panel of non-experts called together by government was presented with information about a new technology – food irradiation – by a range of specialists with differing assessments of the technology's risks. The panel was asked to make its judgment. This in turn had to be defended before the specialists. Thus a consensus was arrived at and presented to the government.[17]

There should be proper debates about whether pesticides are needed, how to ensure safety and how to control environmental impact. To exclude those people at risk when deciding about risk is inherently biased. It is hard to see how it can be justified when trying to make a balanced assessment. Decisions are made by experts who are neither accountable to the public nor representative of many public interests. Since assessment judges the balance between risk and use, why are those who face the risk – consumers, workers farmers – not included?

What Standards Are There?

The debate about pesticides always ends up with standards. Why? Who sets them? Whose are the best? Are they tough enough? Why do they differ? Are they worth the paper they are written on? Standards are worked out in different ways, but there are common elements. In this chapter we look at standards for food, water and workplace exposure, and ask whether pesticides should be assessed for whether or not they are needed. We explain the terms used and assumptions made by those who set the standards. These standards will cause more and more quarrels as countries and trading blocs around the world vie with each other. (See Chapter 11 for the latest details.)

Food

NOELS, ADIs and MRLs

The *No-Observed-Effect Level* (NOEL) is the maximum dose at which specific treatment-related effects do not occur. Please remember that the whole purpose of setting standards is to ensure that no harmful effects take place – there is no conspiracy to kill you deliberately! The NOEL is divided by a safety factor to calculate the *Acceptable Daily Intake* (ADI).

Usually the safety factor is set at the figure 100. This is based on allowing 10 times for variation between animal and humans and 10 times for possible variation among between individual humans. $10 \times 10 = 100$. Different safety factors may be set for different foods or chemicals. According to the Advisory Committee on Pesticides, 'other values may be considered more appropriate depending on the nature of the reported effect'.[1] This is a recognition by regulatory authorities that they have more confidence in some data than in others.

The Acceptable Daily Intake (ADI) is defined as 'the amount of chemical, expressed on a bodyweight basis, which can be consumed daily in the diet over a whole lifetime in the practical certainty on the basis of all the known facts that no harm will result'. The veil is beginning to lift. Here is a committee you have probably never heard of which sets standards for your pesticide safety and residue intake. Note the word 'acceptable'.

The ADI is usually set by the World Health Organisation and harmonised internationally by a committee with a Latin name – the Codex Alimentarius Commission of the United Nations, usually referred to as Codex Alimentarius or just Codex (see also p. 99). Consumer representatives are allowed on to Codex, but only as part of their government's delegation, and if invited often have to pay their own expenses. Not many go. Some consumer groups think this is undemocratic. We believe there should be a public register of who pays the expenses. The Codex Alimentarius also sets the *Maximum Residue Levels* (MRLs). These refer to the maximum residue allowed for any particular pesticide in the food as it leaves the farm gate. MRLs are calculated by working out the maximum quantity of the given produce that anybody could possibly consume to ensure that it would not exceed the ADI.

When MRLs were initiated they were not intended to be safety levels. That is why when the UK 1988–9 residue tests showed that 6% of fruit and vegetables or 0.4% of milk, bread and potato samples had residues above their MRLs, quick as a flash MAFF said it was no problem. Why call it a MRL, then? The breach of a MRL should trigger a toughening up of enforcement and monitoring.

If the MRL does exceed the ADI, what then? To us on the outside there often seems to be a shuffling of papers, a bit of recalculating, and an 'ahem' as if to say: 'Well, consumers wouldn't eat at this level for ever . . . and even if they did, not everything in their diet would contain that particular pesticide . . . and anyway the stuff isn't eaten raw . . . and of course not *all* parts of the plant are eaten. . . .' And so on. Hey presto – out pops a reassessment set at a level at which the MRL would not lead to the ADI being exceeded!

Under pressure from the European Commission, the MRL has now been converted into a legal limit and is used for statutory control of each pesticide. All well and good. The trouble is that there are MRLs for only 62 pesticides. When the Pesticides (Maximum Residue Levels in Food) Regulations 1988 came into force in January 1989 they omitted 31 which had been proposed the previous April! Even worse, the final regulations produced a further 24 MRLs which were less stringent than the original proposals.

So please understand our touch of cynicism about how MRLs

are set. It looks horribly as though standards are dropped to suit circumstances. Is the public interest losing out again?

The Fourth Hurdle: do we really need that pesticide?

What do regulatory committees do? They evaluate safety, efficiency and effectiveness. They do *not* ask an increasingly sensitive question: is this pesticide necessary? The issue of need is now a hot potato in all food regulatory circles. Need is sometimes referred to as the 'Fourth Hurdle', after safety, efficiency and effectiveness.

Biotechnology and drug companies, for instance, are extremely hostile to the Fourth Hurdle. They argue that assessment of need would be a block on investment in new products and technologies. We don't share their hostility. We think that the argument about the Fourth Hurdle is really about democracy. In truth, companies already decide about need – *their* need. They decide about whether a product or pesticide is necessary, but their criteria for need are more to do with sales than with social or environmental impact.

Water

MACs

The European Commission has set *Maximum Admissable Concentrations* (MACs) for pesticides in drinking water. The directive, or European Community law, came into effect in 1985. After some huffing and puffing, the UK's laws came into line when the Water Supply (Water Quality) Regulations 1989 were passed.

The MAC for pesticides is 0.1 part per billion for individual pesticides and 0.5 parts per billion for total pesticides. The British Agrochemicals Association does not like the approach of considering all pesticides as the same and lumping them together! It argues that 'it is sensible to use the extensive data base available to calculate individual MACs for each pesticide'. You can work out what we think of that from Chapter 9.

Standards for pesticide residues in water are tough, no doubt about it. Critics say this was a political, not a scientific, decision. We note with interest, however, that the 1990 British Medical Association report laid as much stress on its concerns about water residues as on anything else.

If you are not already bewildered by the number of different systems and acronyms – NOELs, MRLs, ADIs and MACs – try to be patient. Unfortunately there are more to come!

At work

OELs, MELs, TLVs and IRISes

Occupational Exposure Limits (OELs) have been set for about 30 pesticides. None have *Maximum Exposure Limits* (MELs). There is no list which separates out pesticides from all the other chemicals in the complete OEL list – despite requests from trades unions for the authorities to do so. The OELs are now backed up legally by the Control of Substances Hazardous to Health Regulations 1988 (COSHH). These OELs are based largely on older *Threshold Limit Values* (TLVs).

Most TLVs were set 30 or so years ago and have not been revised. They refer to:

the airborne concentrations under which it is believed nearly all workers may be repeatedly exposed day after day without adverse effect. Because of wide variation in individual susceptibility a small percentage of workers may experience discomfort from substance at or below the limit. A smaller percentage may be affected more seriously . . . by development of an occupational illness.[2]

The TLVs were set by the American Conference of Government Industrial Hygienists (ACGIH), with no clear method of assessment. You can look up how decisions were made in the 'Documentation of Threshold Limit Values'.[3] You will find that the level for Lindane of 0.5 milligrams per cubic metre (mg/m^3) 'is believed to be sufficiently low to prevent central nervous effects'. This judgment refers to and is based on three studies of animals exposed for between one and two years, and three studies on people exposed for between three days and two years.

One of the human studies found that technical Lindane at 40 mg per person per day for 14 days produced diarrhoea, vertigo and headache. Another study found that most patients tolerated 135 mg a day for three days, but one patient became epileptic. Of 37 workers exposed to undisclosed levels over two years, three had serious EEG (brain impulse) disturbances, and 14 suffered minor symptoms. These are the only studies quoted to explain the 0.5 mg/m^3 level that is still in place and that is supposed to protect for a lifetime exposure.[4]

The 'Documentation', produced in 1988, also states that 'some authorities believe that man is more sensitive to Lindane than animals'. The ACGIH 'Committee recommends the deletion of the Short Term Exposure Limit (STEL) until additional toxicological data and industrial hygiene experience become available'. In 1990, it was unchanged! Likewise in 1986 the TUC asked the HSE to reduce the OELs of Aldrin, Chlordane, Dieldrin, Picloram, Pentachlorophenol and 2,4,5-T to that of Captofol, i.e. 0.1 mg/m³. In the 1990 list the levels remain at between .25 and 10 mg/m³.

In the USA there is another set of proposed levels called *Integrated Risk Information System* (IRIS). IRIS exposure guidelines and risk estimates were developed by the US Environment Protection Agency for chronic exposure. This system is based on current toxicity data and on a clear, consistent method of assessment. The method of assessment explains assumptions and how calculations are made and how they are standard for all chemicals. IRIS standards do not include considerations of feasibility and cost.[5]

IRIS uses a measure called the WAC, the maximum time-weighted average air concentration which is expected to cause no adverse effects in humans over a 40-year working lifetime. WACs are calculated for non-carcinogens based on safe levels. Carcinogens are allowed to pose a risk of one in a million, as they cannot be considered safe at any level. The RF is the risk factor, the amount the OEL exceeds the level derived for IRIS. For instance, in the table below the OEL level for Pentachlorophenol allows 1.9 times more contamination than the WAC level.

Let's look at how the systems compare in practice.

Comparison of Five Problem Pesticides on Three Systems for Setting Pesticide Exposure Occupational Standards

Pesticide	OEL (mg/m³)	WAC (mg/m³)	RF(OEL/WAC)
Carbaryl	5	0.9	5.6
Chlordane	0.5	$1.42 \times 10(-5)$	35,211
Heptachlor	0.5	$4.046 \times 10(-6)$	123,579
Lindane	0.5	0.0027	185
Pentachlorophenol	0.5	0.27	1.9

OEL = occupational exposure limit
WAC = weighted average air concentration
RF = risk factor (by which OEL exceeds the level derived for the US IRIS system)

At home

Standards are set for pesticides in food, in water and at work, but none for the general or domestic environment. In certain countries, household chemicals and pollution in homes are already recognised as a serious concern. A major test case on a wood preservative known to be carcinogenic, but sold by a US company in Germany, was begun in the Frankfurt courts in 1990.[6] The case was lost. Environmental and consumer researchers in Australia are now monitoring pesticide levels in people's homes.[7] Australians are very aware of domestic pollution, partly because for years pesticides have been liberally used against termites in buildings, and against the bugs and spores that thrive in the warm, dry climate.[8]

If there are standards for food, water and the workplace, why not the home? We say:

- any level of pesticide exposure or intake can be significant. A small residue does not mean it is insignificant
- standards don't operate in a vacuum. Consumer, farmers, workers, doctors, in fact anyone with a healthy concern about pesticides should be wary of accepting standards at face value
- it is time to open up the committees which set the standards. There are no consumers, environmentalists or worker representatives on the UK committees setting or monitoring standards. What have the committees got to hide?
- an assessment of whether a pesticide is needed should be included in the judgments that committees are asked to make – they should not just be restricted to safety, efficacy and effectiveness
- there should be a consistent method of assessment for wherever you are exposed to pesticides.

Who Controls Pesticides Nationally and Internationally?

Because pesticides are found all over the place, there are all sorts of regulatory bodies involved. Some of these bodies determine how the pesticide comes into the environment. Others control them once they have been released. Obviously their roles overlap, leading to confusion and rivalries. There are three levels of control which you should be aware of: world, European and UK. In the 1990s there will be increasing argument about who has the right to impose controls, and whose controls are the best, for whom. Consumers and industry, for example, may have different understandings of the definition of good control.

World controls

The international trade in pesticides is covered by two United Nations conventions. The UN Food and Agriculture Organisation (FAO) publishes *The International Code of Conduct on the Use and Distribution of Pesticides* (known as the *Code*). The UN Environment Programme has produced *The London Guidelines for the Exchange of Information on Chemicals in International Trade* (known as the *London Guidelines*). World standards for pesticides are important for the General Agreement on Tariffs and Trade (GATT). The effects of the new GATT rules are discussed at the end of this chapter.

The FAO *Code* covers the management and use of pesticides, their availability, distribution and trade. It also deals with information exchange, and with recommendations for labelling, packaging, storage, disposal and advertising. The *Code* is addressed to governments and industry, and also to Non Governmental Organisations (NGOs), trade unions and environmental agencies. The *London Guidelines* are so called after the meeting that produced them and which took place in London.

They are only addressed to governments, but deal with pesticides in the context of other toxic or industrial chemicals. They are mainly concerned with the international exchange of information.

The FAO *Code* is no more than a text arrived at by representatives of governments, by consensus. It sets out the ideals to be aimed at. It does not, however, have any provision for monitoring or enforcement; neither are these tasks undertaken in any systematic way by either governments or industry. In order to remedy these omissions, and to draw attention to their effects, NGOs in both the developed countries and the Third World have taken on the task of monitoring and surveying the effectiveness of the *Code*.

Both the FAO *Code* and the *London Guidelines* have incorporated the Prior Informed Consent (PIC) provisions. The object of PIC is to try to reduce the availability in Third World countries of hazardous pesticides that have been banned, withdrawn or severely restricted in their country of origin. PIC means that no such pesticide should be exported, particularly to a Third World country, unless that importing country has given express consent to the import. The consent should be given before the shipment, and should be informed consent: the importing country should have been given full data on the health, safety and environmental effects of the pesticide.

The Pesticides Trust has produced for PAN a report on the monitoring and implementation of the FAO *Code*.[1] It provides evidence of the wide availability in developing countries of acutely hazardous pesticides that are responsible for documented cases of human poisoning and environmental damage. The report describes the background to pesticide use in the 17 countries concerned:

- in Africa and the Indian Ocean: Egypt, Mauritius, Senegal, Sudan and Zaire
- in Asia and the Pacific: Bangladesh, India, Indonesia, Malaysia, Papua New Guinea, the Philippines and South Korea
- in Latin America: Colombia, Costa Rica, Ecuador, Mexico and Paraguay.

For each country it lists the pesticides which are widely used, together with the particular ones that are causing health or environmental problems.

Some four years after the introduction of the FAO *Code*, breaches were routine and ignored. They commonly occur in advertising, packaging and repackaging, conditions of use, unrestricted availability, the way products were sold, quality control and container disposal. All NGOs reported that, if legislation existed in their countries, it was poorly enforced. Importing governments frequently do not have the resources to monitor pesticide use and the observance of the *Code* within their borders. The pesticide industry concept of 'product stewardship' is not the most effective means of enforcement and monitoring.

European controls

The European Commission, meanwhile, has increasing control over pesticide use. It is committed to removing barriers to trade and to protecting people by proposing common rules on health and safety. This is part of the 1992 agreement, which followed the passing of the Single European Act in the mid-1980s. In Article 100a of that Act, the European Commission (EC) has a strong commitment to set high levels of environmental, consumer and public health protection, and to get European member states to implement them.

There are EC directives (laws) which:

● set residue limits for certain pesticides
● prohibit the placing on the market of certain plant-protective products.

We now outline these directives and also point to some future directives that are currently being discussed. These might go further than those which are already law – it depends how strong and successful the lobbies are.

European residue limits

The Residues Directive (76/895/EEC as amended) sets out Maximum Residue Levels (MRLs – see Chapter 10) for over 40 pesticides and for the fruits and vegetables on which they may be used. The directive allowed national governments to set different MRLs – or none at all. If a member state considers the maximum level too high, it can reduce the level in its own territory but must inform the Commission of its action. The Council or Commission will decide whether to amend the MRL.

The Residues in Fruit and Vegetables Directive (90/642/EEC) was adopted by the Council of Ministers on 27 November 1990.

The framework Directive: extends the range of products for which MRLs can be set to cover potatoes, pulses, oilseeds, tea and hops and classifies products more thoroughly than before; it provides for the setting of mandatory MRLs rather than the optional levels set in the existing Fruit and Vegetable Directive; it applies the MRLs to the specified plant products when used as animal feeding-stuffs. The Commission will review after a 3 year period whether these MRLs would more appropriately be placed within the EC Directive regulating undesirable substances in animal feeds; it applies the MRLs to exports. An earlier proposal that post-harvest pesticide treatments be labelled was deleted. Instead a statement was made in the Council minutes asking the Commission to carry out a study of all aspects of pesticide treatment labelling and to submit any appropriate proposals as soon as possible and in any event not later than 30 June 1991 in order to enable the Council to take a decision by 31 December 1991.

The Cereals Directive (86/362/EEC) came into force in 1988. It applies to a range of cereals for human consumption. Member states must ensure that cereal products are not a danger to health due to pesticide residues. Cereals below the limit are allowed free movement within the 12 member states of the European Community. Member states must make sure the products comply, and reports on how this is done have to be made each year. Member states can use their previous methods. As with the Residues Directive, member states can introduce their own lower levels of MRLs. They can also authorise higher levels if they can show that these new levels will be down by the time the product gets to the consumer. MRLs can be amended by majority vote. New ones require the unanimous vote of the Council.

The Animal Origin Directive (86/363) sets MRLs for certain pesticides in or on meat, poultry, dairy products and processed meats. It works in much the same way as the other directives, and includes levels for Aldrin, Dieldrin, Chlordane and DDT.

Directives also limit the amount of *pesticides residues in water;* (see Chapter 10.)

European controls on pesticide products

Common sense suggests that the best way to prevent problems is to stop them at source. That means preventing the wrong kind of products getting to market or being used. Clearly, this admirable public health principle has gone a little awry over pesticides. But the European Commission is aware of this, and Brussels is increasingly the location of some fairly bitter arguments about pesticide products. Here are the current relevant directives.

Under the *Prohibition of Certain Plant Protection Products* (79/117/EEC), member states have to ensure that certain plant protection products are neither placed on the market nor used. Chlordane, Dieldrin, HCH with less than 99% gamma isomer, and Heptachlor are banned under this directive. Member states are allowed 'derogations' (in Euro-speak that means exemptions) for mercury compounds, Aldrin, DDT, Endrin and Heptachlor. The specified reasons for this are set out in the directive (and amended in 83/131/EEC). Member states are required to carry out a review every two years.

A future directive proposes to go further. The draft Plant Protection Products Directive (Registration Directive) (Com {89}34 final) requires *prior* authorisation according to Community Rules of all plant protection products. If this directive gets on to the statute books, a pesticide will only be authorised if it is on the EC permitted list, if it is sufficiently effective, if it has no unacceptable effect on plants, if it has no harmful effect on human or animal health and if it has 'no unacceptable influence' on the environment.[2] Quite a tall order.

The right of Europe to set its own standards could be undermined by the General Agreement on Tariffs and Trade (see p. 98).

UK Controls

Until the European directive mentioned above comes on stream, a member state such as the UK still has control over what comes on to the market, the MRLs and their monitoring.

The UK government of the 1980s made full use of this autonomy and of any other areas where flexibility was allowed.

These decisions are made by six Ministries, of which Ministry of Agriculture, Fisheries and Food (MAFF) is one. Under the Food and Environment Protection Act 1985 (see p. 114), the Advisory Committee on Pesticides (ACP) was established. The Ministers of the Environment, Health and so on are required to consult the ACP when they contemplate making pesticide regulations, making approvals, suspensions or revocations, and setting conditions to which approvals may be subject.

At the time of writing, the ACP consists of the following people. Chair: Professor C. Berry, Director of the Pathological Institute at the London Hospital; deputy chair: Professor G. Sagar, Vice Principal of the University College of North Wales; Dr B. Buckley, Senior Clinical Lecturer Birmingham University; Dr T. Aw, Senior Lecturer University of Birmingham; Dr A.D. Carter, Senior Research Officer, Soil and Land Research Centre; Professor A. Dayan, Director of the DHSS Toxicology Laboratory, London University; Professor E. Jones, Professor of Microbiology, Portsmouth Polytechnic; J. Leahy, Regional Scientist, Severn Trent Water Authority; Professor D. Lee, Dept of Agricultural Zoology, Leeds University; Dr A. T. Proudfoot, Director of the Scottish Poisons Information Bureau; R. S. Tayler, Senior Lecturer, Dept of Agriculture, Reading University; Dr K. Woods, Senior Lecturer in Clinical Pharmacology, Leicester University.[3] Spot the consumer or worker representative in that list!

The Advisory Committee on Pesticides is supported by a network of subsidiary bodies, comprising its scientific sub-committee, four panels and two working parties. The panels consider application technologies, label and container design, medical and toxicological aspects, and environmental matters. These panels are open to a limited number of consumer and trade union representatives. The working parties deal with pesticide residue levels and pesticide usage surveys. (See Chapter 12 for the ways in which the ACP controls approvals, suspensions and revocations. See Chapter 10 and Table 3 for information on how the ACP has controlled residue levels.)

In May 1990 the Pesticide Trust and Parents for Safe Food wrote Professor Berry a long letter concerning the functioning of his committee. We enclosed a copy of the European Consumers' Pesticide Charter – 10 pages of specific and reasonable recommendations. One point we made was to suggest that some fresh blood might be added to the ACP by including independent. We are still waiting for a reply. Perhaps this is due to lack of resources. In a truly British way, the ACP members are given hardly any resources. For example, one member informed us that he received only a tiny bit of help with his telephone bill.[4] This is scandalous. These people need support and assistance.

And what about the exemptions ('derogations') to European directives that the UK has taken? After much delay, DDT was completely banned in the UK in October 1984. Aldrin was eventually banned 'with immediate effect' in May 1989. Derogations that the UK decided it does not need are some of those for mercury compounds and Endrin on certain propagating stock.

While six Ministries have control over what comes on to the market and how it should be sold, the government Health and Safety Commission (HSC) has certain responsibilities. Under the Food and Environmental Protection Act (FEPA) and the Control of Pesticide Regulations 1986, the Minister of Agriculture consults the HSC regarding worker health and safety before delivering an opinion on certain recommendations from the ACP. This is done through the Chemicals in Agriculture Working Party (CHEMAG) and its sub-committee CHEMSAP. CHEMSAP members can only receive confidential information verbally. They have met only once in three years.

The Health and Safety Commission also has control over the policing of the pesticide laws. When the FEPA was introduced to Parliament, it was proposed that another 18 agricultural inspectors would be needed to police the new regulations. The money was never allocated, despite protests from the civil service.

How effective is this UK system of control?

In the period January 1989 to March 1990 29 cases were brought and 15 prosecutions were made. A total of 1,184 enforcement notices were served. Of these, 812 were to do with storage, 296 with use and 76 with the sale and supply of pesticides.[5]

The Health and Safety Commission and the Health and Safety Executive are tripartite. That means they include representatives from industry, trade unions and other government authorities. They are responsible for virtually all matters of health and safety, yet have little control over the pesticide laws drawn up by MAFF. These anomalies were highlighted in one episode during the introduction of the Control of Substances Hazardous to Health Regulations (COSHH – see p. 117). Attempts were made to put pesticides outside the scope of this law. This was odd. By definition, a pesticide *must* be a 'substance hazardous to health'! Representatives from the National Farmers' Union then complained that two sets of laws, FEPA/COPR and COSHH (Chapter 13) were too much. It was pointed out that the two laws covered different areas: one covered the environment, while the other covered workers. But sufficient confusion had already been created; MAFF and HSE got together to produce *Pesticides: Code of Practice for the Safe Use of Pesticides on Farms and Holdings*.

Independent research

Who makes all the decisions? Who decides what names will sit on their committees? How do they make decisions? The government appoints ACP members, but MAFF services them. Obviously powerful pressures are involved in the world of pesticides.

The general approach, common to many countries including the UK, is a relic from the age of chivalry and, perhaps, 'gentlemen scientists'. Decisions are left to 'scientific experts'. The public are told that this is because matters are so complex that they would not understand. Other reasons given are

● to protect confidential information
● the 'experts' can be relied on to be independent
● and there is little evidence that they have made any big mistakes up to now.

We do not think much of this scientific approach.

The government's approach to research in general is to distinguish between pure or academic work and work which is 'near market', i.e. which has economic implications. Where work in the latter category is concerned, industry is being encouraged to pay for it. Although we agree that industry *is* heavily affected by such research, we are very unhappy that a creeping self-regulation is emerging. Public and environmental health deserve better.

So we wonder whether any expert committee can be really independent of the chemical manufacturers. The whole science of pesticides, the whole structure of thinking, is dominated by the manufacturers. They decide what pesticides are investigated and what direction the research shall take. They have all the money. They make the investment for future products (it can cost up to £30 million to get a single product on to the market). And they give grants to academic institutions, who are more than grateful to receive them. With few other current sources of funding, academic institutions are even less likely to be critical – only with alternative sources could they research alternative methods of control. In practice it is hard to divide industrial research and academic research. The researchers all go to the same conferences, they all get on well, and they exchange a lot of information – but not all. How can anybody or any committee claim to be independent? We do not accuse the scientists of being 'bought off'. We are sure that they are conscientious and well-meaning. But how can they meet the need for independent science with such meagre resources?

Big changes

What is going on at world level? World trade is regulated by a body called the General Agreement on Tariffs and Trade (GATT). GATT comprises both the set of negotiated agreements and the international forum that enforces them, covering some US$ 3 trillion of trade annually. It seeks to standardise trade practice in order that there can be free trade. The agreements are periodically reviewed in a multi-year session called a 'round'; there have been seven rounds since GATT was set up in 1947.

In December 1990 a new GATT – the Uruguay Round – failed to get agreed. One of the thorniest issues, argued over for four years, has been agriculture and food trade.[6] The world's most powerful trading blocs, the USA and the European Community, are both food exporters. Both have different systems of subsidy and support for their farmers. Both accuse each other of providing unfair subsidies. Both want a bigger slice of the action. There was also much conflict in this round over the wide differences between each nation's food safety and environmental standards.

The GATT Uruguay Round restarted in 1991. Plans propose to give more power over pesticides, all food standards, and health and quality issues to a body called the Codex Alimentarius (already mentioned in Chapter 10), administered by the UN Food and Agriculture Organisation and co-financed by the World Health Organisation. Codex Alimentarius is unelected and accountable only to governments. It is made up of around 130 government delegations. These have included executives from chemical and food companies. There is only one consumer body and no worker or environmental representative on this huge and increasingly important committee. The US delegation to Codex has included representatives from government departments like Agriculture and the FDA. It has also had representatives from Nestlé, Coca-Cola, Pepsi, Hershey Foods, Smith Kline and Kraft, together with people representing the Grocery Manufacturers of America, the Association of Cereal Chemists and the National Food Processors. The US delegation is presided over by a White House appointee from the US Department of Agriculture, currently Dr Lester Crawford.[7]

Corporations are lobbying for new GATT rules which would limit the right of nations to set stricter standards and to allow other governments to pre-empt pesticide and food laws. Under the regulations aimed at by the industry's giants 'scientific evidence' would be the only consideration in human health and environmental regulations applied to imports. No social, economic, religious or cultural concerns would get a look in. (See how this conflicts with EC proposals outlined on p. 92)

In the USA there is growing concern that the country's own standards will be watered down by the new GATT system, (GATT's *existing* standards are often lower than the best in the world). Here are the Codex pesticide MRLs on foods which are weaker than US EPA tolerances. Let's compare the Codex MRLs with the US EPA tolerances.

Crop	Pesticide	Increase
Carrots	Benomyl	25×
Apples	Permethrin	40×
Strawberries	Lindane	3×
Potatoes	Diazinon	5×
Bananas	Aldicarb	1.6×

And here are some examples of Codex pesticide MRLs on foods which are weaker than US EPA action thresholds.

Crop	Pesticide	Increase
Broccoli	Heptachlor	5×
Grapes	DDT	20×
Milk	Endrin	3×
Peaches	Dieldrin	50×
Lettuce	Aldrin	3.3×

Source: Bill Barclay, Greenpeace USA.[8]

The new powers being proposed for Codex Alimentarius have yet to be finalised as this book goes to press. But already it is clear that the process by which environmentalists, health and safety specialists and consumer organisations have pushed and shoved to achieve tougher standards of protection will be harder. Gradually, pressure has forced improvements in controls – now in this country, now in that, now for this pesticide, now for that. This process of working to the best standards will be impossible if Codex Alimentarius becomes the body with huge and increased world influence on standards. Codex should be completely overhauled.

The new GATT round could increase the control of powerful nations and big business over the production of world food. The right of national governments and regional legislatures to implement and protect themselves with environmental and health protection measures will be seriously compromised.[8] The right of smaller, less powerful companies and countries to influence the trade will also be decreased. As the Research and Development Director of ICI said in 1990: 'More stringent registration requirements provide higher entry barriers to prospective new entrants.'[9] We predict that registration requirements may go up, but standards for workers and food may not.

Pesticide of the Month – or Last Year's Fashion?

C hapter 11 explained how the Advisory Committee on Pesticides (ACP) is responsible for advising UK ministers on what pesticides should be available. In this chapter we look at the lessons to be learned from watching the comings and goings of the government's approved list of pesticides, published annually.[1] One moment a pesticide is considered all right by the powers that be; the next moment it is off the list. Sometimes the permitted use for a pesticide changes. One moment a pesticide is approved for agricultural, the next only for non-agricultural use, with little or no explanation as to what the differences in risk might be. What is going on here? Answering this question can throw some light on how judgments about danger are made, as discussed in Chapter 9.

Reclassifying the pesticides

Since the Control of Pesticide Regulations 1986 (COPR) were passed, the total number of pesticide formulations in the MAFF-approved list has dropped by about 100. The total has grown by around 150, however, if the new 'Amateur Products' list is included. There were about 1,400 mixtures in 1990. There is also a new HSE list for non-agricultural uses. In addition, some agrochemicals have swapped uses. For instance, tar oils have become a herbicide, as has copper sulphate! Ditalimfos used to be classified as an insecticide, but now it is a fungicide. Ziram used to control fungi, and now controls vertebrates.[2]

Arrivals . . .

Since the implementation of the Control of Pesticides Regulations, the Pesticides Registration Department of the Ministry's laboratory at Harpenden in Hertfordshire is where all applications for approval are received. The application is then put into an appropriate assessment procedure – 'fast', 'normal' or 'committee'.

'Fast' means the application does not require re-evaluation, but only minor changes. 'Normal' includes those requiring experimental permits, minor extensions and the evaluation of supporting data. Approval is granted after government departments have agreed the proposed use and accepted the label. All new active ingredients and major extensions of existing products are put into the 'committee' procedure. Supporting data is provided, as indicated by *Data Requirements for Approval under COPR 1986*. The data is first evaluated 'in-house' – the committee produces a report for the advisers on the scientific sub-committee. The evaluation assesses the supporting data regarding the fate and behaviour of the compound in soil and water, in crops and animals.

For some time, MAFF's Data Evaluation Unit at Harpenden was understaffed, underpaid and housed in portable, temporary offices. Files were stored in a different part of the country. Staff morale was low – understandably so, in view of the higher salaries and better working conditions that similarly experienced scientists would enjoy in industry. In April 1990 MAFF announced that the staff of scientists who conduct the reviews was to be increased. Suddenly, MAFF intended to 'treble' its capacity for the safety testing of pesticides, a welcome if belated recognition of previous inadequacies.

Toxicity is assessed by the Scientific Sub-Committee of the Advisory Committee on Pesticides (ACP). So are the likely levels of exposure. We do not know *how* the assessment is made (see Chapter 9) because the process is secret and there are no public interest representatives on the committee.

The Scientific Sub-Committee reaches a consensus view before making a recommendation to the ACP. The ACP may amend the recommendation before it goes to six government departments – it may recommend full approval, experimental permits, or provisional approval for further evaluation.

The HSC is supposed to be consulted in all cases where the full approval of a pesticide is recommended and where significant health and safety aspects apply to pesticides subject to review, revocation or final approvals. The responsible body within the HSC is the CHEMSAP Working Group. Trade Union representatives on that body complain that they have only been consulted on seven new active ingredients and ignored entirely in the review process (see p.110). CHEMSAP members can only be given commercially sensitive information on a verbal basis at meetings. They held just one meeting between 1986 and 1990.

Applicants agree to the public release of the evaluation before the approval is granted. Notices announcing approval are published in the *London Gazette,* the *Edinburgh Gazette* and the Pesticides Register. The HSE has similar procedures for applications submitted to it. There is also a system for rapid approval of imported pesticide products, identical to products already approved under COPR, from the European Community to the UK.

The *Pesticide Register* is the monthly list of UK approvals. It is produced by MAFF and HSE and published by HMSO at a price of (in 1990) £30 a year. Since the introduction of COPR in 1986, the following active ingredients have gained approval:

- Flocoumafen, a rodenticide
- Quizalofop-Ethyl, a systematic herbicide
- Cyfluthrin, a contact insecticide
- Ethoprophos, a nematicide and soil insecticide
- Myclobutanil, a triazole fungicide
- DPX M6316
- Azaconazole
- oxine-copper
- Fluazifop-p-butyl
- Flusilazole
- Bifethrin
- IPBC
- Tributyltin naphthenate
- Fenoxaprop-p-ethyl
- Fenoxaprop-ethyl
- HOE 070542 triazole coformulant
- PP321
- Cyhalothrin
- Fenpropathrin

For this book we have done a check, and found the following new names in the MAFF-approved list since 1986:

- Chlorotoluron
- Diflufenican
- dodecylbenzyl trimethyl ammonium chloride
- Imazamethabenz
- Pentanochlor
- Thifensulfuron-methyl
- Flusilazole
- 8-Hydroxyquinoline
- nonylphenoxypoly(ethyleneoxy)ethanol-iodine complex
- Pencycuron
- sodium carbonate, sodium metabisulphite and sodium propionate
- Bifenthrin
- Etrimfos
- Diphacinone

We have included these lists in this book to show how fluid the comings and goings of approval are. The *Pesticides Register* now lists new approvals.[3] But unless you get an up-to-date copy of this publication every month, it is very hard to find out which pesticides have been approved.

Since January 1985, only 35 applications for approval of new pesticide products have been processed. As of the end of August 1990, there were 45 applications in the queue waiting to be processed for approval or rejection.[4]

. . . And departures

Aldrin and Dieldrin

Although it was nearly 20 years after some countries had curtailed their use, before the decision was taken to replace 'the drins', certain uses were still permitted until 1988. The ACP reviewed them and considered that any concentration in the environment was undesirable. It also considered that the European Community limits were not strict enough. On this basis the ACP recommended that all products containing Dieldrin should be revoked from March 1989. But it considered that Aldrin would need a phasing-out period, so the revocation for sale, supply and advertisement was delayed until the end of 1990. Approvals for use and storage would continue till the end of December 1992! However, the next year the ACP was told about very high levels of Dieldrin in a Cornish river. Dieldrin degrades to Aldrin, so the ACP recommended, and the Minister accepted, an immediate ban on Aldrin products.

The main manufacturers organised a reverse chain disposal scheme (where the manufacturer goes back and collects existing stocks), at no cost to the growers. Mind you, because of the 1988 warning – not to mention the decision 20 years before – there were not many stocks left.

Binapacryl

Approval revoked (see also Dinoseb).

Captan

Approvals limited.

Captafol

In 1989 the ACP banned Captafol on edible crops because it was found to be carcinogenic, although not non-genotoxic (See Chapter 6). Subject to further review, Captafol could still be used as a paint for canker control and as a fungicide when chipping narcissus bulbs. But it was suspended completely in 1990.

Chlordane

Despite the EC directive to phase out all uses of Chlordane, nearly 50% of UK sales of weedkillers in the mid-1980s contained it. The ACP recommended in 1988 that all Chlordane uses should be revoked from the end of 1988 – except for earthworm control on fine turf used for sport. It considered that the withdrawal 'would result in a considerable increase in manpower and materials without achieving equivalent standards' (in other words, you would need more people applying more – though different – chemicals, to achieve a poorer result). Sales and supply for this purpose are revoked from the end of 1990 and approvals for use and storage revoked from the end of 1992. This advance warning of a ban is almost an invitation to get stocks in!

Cyhexatin

In 1987 the ACP recommended, with agreement from CHEMAG, that pesticide products containing Cyhexatin should be revoked. This was after skin studies on rabbits had shown teratogenic effects, thus posing risks to pregnant women. Sales were allowed until May 1988. This was not a 'reverse chain dispersal', thus allowing withdrawals and dispersal of stocks.

Dinoseb and related Dinitrophenols (except DNOC – see below)

Dinoseb was a 'specified substance' because between 1948 and 1951 eight agricultural workers died of DNOC poisoning (a closely related compound). In October 1986 the US EPA announced an immediate ban on Dinoseb because of evidence from new teratogenic studies that there was a risk of birth defects to workers. Based on three birth defects in three animal species exposed to Dinoseb, in December 1986 the ACP suspended approval for Dinoseb, Dinoseb-acetate, Dinoseb-amine, Binapacryl and Dinoterb. This marked the first time that ministers had decided upon a rapid 'emergency' action to ban an approved pesticide. It showed they could do it if they wanted to.

A few months later suspension was relaxed for Dinoseb sprayed on Glen Cova raspberries, a famous variety, in Scotland – not as a herbicide but to control cane vigour. The ban was lifted for a year, provided the chemical was applied only by named male operators. But the local agricultural workers' union claimed it had the results of unpublished tests done by two major international chemical companies in the 1970s, which proved that Dinoseb was a testicular toxin and caused sterility in men exposed even to very small amounts. The Health and Safety Executive also opposed the relaxation of the ban.

DNOC

In 1988, the scientific sub-committee of the ACP asked for teratogenic data from the data holders. None was provided. Given the identified risks, the sub-committee recommended that DNOC be revoked. According to the ACP, this is the first time that the question of deficient data had been addressed. The ACP recommended a phased revocation of approvals. Sale and supply would continue to the end of 1988, and storage and use for another year only.

Ioxynil and Bromoxynil

Following an extensive review caused by new evidence about teratogenic, thyroid and male reproductive effects, the ACP found several areas of concern. There was inconsistency in the data on safety margins, and an absence of two major items of data – on skin teratology and residues in onions. They were also concerned that the studies undertaken over the past 20 years had been poorly performed and that the information provided was not complete or up to standard. The ACP recommended that all home and garden uses for Ioxynil should be revoked immediately.

Other approvals for Ioxynil and Bromoxynil were continued on a provisional basis. Alterations were made to the labelling, Maximum Residue Limits, containers, and timing of applications on cereals, grassland, onions and leeks.

Piperonyl butoxide

This one has gone, but we do not know why or how. It is a 'potentiator' of other insecticides, and not really an insecticide in itself, but it was on the MAFF-approved list in 1986. It is not there now, but it is still found in all parts of the world. In the early 1970s it was known to inhibit various cell enzymes and to slow down the process of removing the poisons of pesticides. It had also produced liver cancers in mice studies. We ask:

- has it gone for good?
- have the manufacturers removed it?
- did ACP revoke it?
- or has it just changed its classification?

Other pesticides gone from the approved lists since COPR in 1986 are:

- ammonium thiocyanate
- Bitertanol
- Butoxypropylate
- Cholin Chloride
- CMPP
- Dimefuron
- Di-1-P-Menthene
- Ethoxyquin
- hydrogen cyanide
- Iodophor
- organomercury compounds (EC directive)
- sodium trichloroacetate, sodium nitrate and sodium chloride – the last of these better known as salt!
- soft soap as an insecticide.

Apart from the organomercury compounds, we do not know why these chemicals have disappeared from the approved list.

- Folpet, a fungicide used in anti-fouling paints, has also gone. Although this pesticide had been banned for several years in many countries and is not approved for use in the UK, in 1989 the ACP called for data on it in order to review its use. That call was cancelled in 1990 as 'there are no longer any approved products containing Folpet'.[5] Thus it was never banned from use in the UK – just lost.

The following active ingredients have had their approvals for agricultural use withdrawn 'for commercial reasons'.[6]

- Aluminium silicate
- Anthraquinone
- Barban
- Benzethonium
- Carbophenothion
- Chlorthiamid
- Cufraneb
- 2,4,5-T
- Di-allate
- Ditalimfos
- Fenchlorphos
- Fenoprop
- Flamprop-methyl
- Fluoroacetamide
- Mevinphos

- oxine-copper
- Quinonamid
- sodium carbonate
- 2,4,5-T
- tar acids
- Thionazin
- Thiourea

We wonder what 'for commercial reasons' means.

The Ministry says 'no specific concerns relating to approved use of these active ingredients have been identified. Approvals will continue for a period sufficient to allow existing stocks to be used up and disposed of safely.'[7] Let's now explore the significance of 2,4,5-T being in that list.

The case of 2,4,5-T

This was the chemical used in Agent Orange, which the US Army sprayed over Vietnam. 2,4,5-T is a herbicide and was used to 'defoliate' the Vietnamese countryside. 'Defoliation' is a polite word for the removal of leaves of living plants. This is so that the USA's enemy could be more easily seen. Many Vietnamese suffered birth defects as a result. US war veterans who had handled the pesticide also fathered children with birth defects. The cause was found to be a contaminant of 2,4,5-T called TCDD. TCDD is one of the Dioxin group of chemicals and is one of the most toxic synthetic substances known to humanity.

When one of your authors was starting his doctorate on pesticides and soil life in 1970, he was advised by his supervisor not to work on the impact of 2,4,5-T on soil animals. He was told the herbicide would surely be banned within three years.

In the late 1970s 2,4,5-T became the focus of a major trade union campaign in the UK. Various farm workers complained that their children had birth defects. The then National Union of Agricultural and Allied Workers organised its members and lobbied publicly to have 2,4,5-T banned. The Forestry Commission agreed to suspend its use for one year. When the Ministry's Advisory Committee on Pesticides said 2,4,5-T was safe to use, the Forestry Commission ended its ban. Deputations by the trade unions to the ACP, its sub-committees and politicians, gained some concessions, but failed to get 2,4,5-T banned.[8]

Today, 2,4,5-T has lost its official approval 'for commercial reasons'. We cheer at its demise, but questions remain. Without a formal ban, can 2,4,5-T be manufactured in the UK for sale abroad? Is the removal for commercial reasons to avoid any constraints being put on pesticide marketing by the new United Nations Prior Informed Consent Code (see Chapter 11)? Under this code, countries receiving particular hazardous pesticides will have to be told about any hazards associated with them. Why not write to your MP and ask him/her to ask the Minister why 2,4,5-T is still not banned?

Public pressure works – but don't hold your breath

In 1988 the ACP announced, after considerable pressure, that it was going to review 100 pesticides approved for use. In 1989 an extraordinary alliance of environment, consumer, labour and agrochemical interest groups joined to express concern at the backlog in the review process.[9] The Minister responded by increasing resources for the test laboratories at Harpenden. But the backlog will still take years to clear, and meanwhile the pesticides continue to be used.[10]

Here is a brief summary of some of the more important reviews.

Aldicarb

In the light of suspicions that Aldicarb may affect the immune system, the ACP looked at the new evidence. One study was dismissed and the other demonstrated no effects. Continued approval was given.

Daminozide

Following the withdrawal and threatened ban on Daminozide in the USA, the ACP reviewed its use and gave it the all-clear. The manufacturers withdrew it themselves worldwide. Formally, it can still be used in the UK, so it might still be used.

Dinocap (similar, but not that similar, to Dinoseb and DNOC)

This was reviewed and given provisional approval for two years.

Fentin acetate and fentin hydroxide

Both these carry on provisionally.

Lindane

The HSE is undertaking an extensive review to include questions about carcinogenicity, effects on reproduction and aplastic anaemia.

Sulponyl urea herbicides

These carry on provisionally.

Tecnazene

This is commonly sprayed on potatoes to stop them sprouting when they are stored over the winter. The committee expressed concern over the safety factor of 1500 (see Government definition in Appendix 2, pp 2 and 3). It felt this figure had been chosen arbitrarily and recommended that the company should undertake the necessary mutagenic studies. At the time of going to press the manufacturers had not produced the necessary data. The ACP was duly 'concerned' and making noises about serving enforcement notices.[11]

Read it for yourself

The Ministry now publishes an evaluation document on each pesticide that has received partial or full review. Each costs (in 1990) between £2.75 and £10.50. Copies are available from Lawrence Ridgley, Pesticides Safety Division, Branch Room A, Room 325A, Ergon House, c/o Nobel House, 17 Smith Square, London SW1P 3HX.

You can get them for

- Daminozide
- ethylene bisdithiocarbamates
- tributylin oxide

- Alachlor
- Fenbutatin oxide
- Fentin acetate
- Fentin hydroxide
- Iprodione
- 2-Aminobutane

Plus ça change . . .

Dichlorvos is still approved as a fly-killer. It is also widely used to treat sea-lice in salmon farms. For many years, Dichlorvos has been contentious. It hit the UK press again in 1990. It has the highest acute toxicity of all the 22 dangerous substances listed on the Department of Environment's Red List and has been used on some salmon farms without a product licence for over 10 years.[12] Following attempts to control its use for this purpose, it was reclassified. Instead of being a pesticide it became a veterinary product. This meant that it could only be obtained from the manufacturer on veterinary prescription to treat specific outbreaks. Yet a recent HSE survey found that 'in practice most farms seem to keep large stocks ready for immediate treatment at the first sign of outbreaks'.[13]

The questions that remain

These comings and goings raise more questions than they answer.

- where there is an exception to a ban, how is it that it is usually where the pesticide is used *most*?
- why is Ioxynil too dangerous in the home and garden, yet available for large-scale spraying? About 100 formulations containing Ioxynil or Bromoxynil are in the 1990 approved list. Shouldn't the occupational user get the same level of protection as the home gardener?
- why did it take two years for DNOC to be 'banned' when its close relative was banned immediately two years before? Forty years previously DNOC fatalities had led to action on Dinoseb. Why the delay?
- how is DNOC still in the approved MAFF list in 1990 – the year after it was supposed to be all phased out?

- what does 'banning' mean? Some pesticides have reverse chain collection, while others are left for a few years for collection. Others are allowed back. Some never go away!
- how many other pesticides are waiting for a sudden study showing new risks?

Perhaps you would like to ask your MP – or the Minister direct – for answers to these questions. The Minister can be contacted at the Ministry of Agriculture, Fisheries and Food whose address will be found on p. 235.

What UK Laws Govern Pesticide Use?

Perhaps because pesticides are found in all parts of the environment, the laws governing them can be all over the place too! Here is a summary of the key pieces of legislation in the UK, which can be bought from Her Majesty's Stationery Office (HMSO) or ordered from your library.

At the start of the 1990s two arms of the law cover the main aspects of pesticide use:

- the Control of Pesticide Regulations (COPR) 1986 was drawn up under the Food and Environment Protection Act (FEPA) 1985
- the Control of Substances Hazardous to Health (COSHH) Regulations 1988 was drawn up under the Health and Safety at Work Act (HASAWA) 1974.

FEPA and COPR

FEPA (Food and Environmental Protection Act 1985) gives legal status to the old voluntary Pesticide Safety Precaution Scheme. The Advisory Committee on Pesticides (ACP) is given responsibility for clearance of pesticides. The Act gives ministers powers

- to set Maximum Residue Limits (MRLs)
- to make information more available to the public
- to seize pesticides that do not conform to the law.

COPR (Control of Pesticide Regulations 1986) prohibits the sale, supply, storage or use of any pesticide unless the ministers have given approval for it and a consent to that activity. Conditions of both approval and the consent have to be complied with. Anyone doing so must take 'reasonable precautions' and be competent to do so. For the first time, extensive training of operators in pesticide handling was required. Nobody can mix pesticides or add adjuvants, unless it is a condition of approval. But COPR does *not* apply to pesticides used

- in manufacturing processes
- as insect repellents on your skin
- in adhesive pastes
- in metalworking fluids
- in paint
- in water supply systems or swimming pools
- and to those manufactured solely for export.

Why not?

Under COPR, detailed specifications for aerial spraying are laid down that relate to the notice that must be given, records kept and conditions applying. Breaches of this law can result in seizure of the pesticide, possible removal of the pesticide from the UK, and any other remedial action the ministers consider necessary. Aerial spraying can be very bad news.

No freedom of information

Regulation 8 enables ministers to set guidelines for the disclosure of information collected during the approval and review process. This is set out in the MAFF document *Disclosure of Information: Procedures and Safeguards*. Information may be made available in several ways:

- details of approved products, uses, and initiation of reviews
- ACP Annual Reviews
- details on usage and residue surveys, and from sales
- toxicological information. If you pay, you can have summary evaluation.

Independent researchers are not given access to the vital raw toxicological data. You may be able to obtain more detailed study reports, but to do so you have to state your scientific qualifications, employer's name, source of research funding, commercial interests, reasons for request, reasons why evaluation reports were not enough, and so on. It is not clear, once you have answered that lot, what criteria there are for deciding whether you can have the requested information. There are probably more controls on pesticide information than on the pesticides themselves!

The information that manufacturers have to supply for approvals and labels is set out in *Data Requirements for Approval under COPR 1986*. The data required is much the same as that required voluntarily by the Pesticide Safety Precaution Scheme. The tests are carried out by the manufacturer, and the results assessed by the SSC and then the ACP (see Chapter 9).

COPR also incorporates the Classification, Packaging and Labelling of Dangerous Substances Regulations 1984, which state that pesticides must have proper labels and packaging. These Regulations also serve as a starting point for the assessment of some pesticides under the COSHH regulations.

Enforcement

In the period January 1989 to March 1990, 29 cases were brought and 14 people successfully prosecuted under the COPR Regulations. Total fines amounted to £10,000. Of 1,184 enforcement notices issued, 296 were concerned with use, 76 with supply and 812 with storage of pesticides. In a survey of stores, 60% were not satisfactory and 73% of these had some form of enforcement action taken against them.[1]

In 1990 the head gamekeeper on the Scottish estate of David Heathcoat-Amery MP, then Under Secretary of State at the Department of the Environment at the time of the incident, was prosecuted for endangering wildlife and storing pesticides illegally. The gamekeeper, Mr McGregor, was fined £1,200 for possessing Alphachloralose and Phosdrin, capable of being used to commit offences under the Wildlife and Countryside Act, for storing them incorrectly and for possessing Phosdrin without necessary approval. The offences came to light during a police search following reports of a sheepdog and a buzzard being poisoned. Said Mr Heathcoat-Amery: 'I very much regret these offences. Any wrongdoing was without my knowledge and consent.'[2]

Are there enough 'policemen on the beat' to be effective in promoting safe use? HSE inspectors are not able to visit many farms very often. Because of the shortage of inspectors and resources, a small farm may not be visited more than once every 28 years! In 1989 a coalition of environment, consumer, trade union and agrochemical interests called for the number of health and safety inspectors in the Agricultural Inspectorate to be increased substantially.

The Institution of Professionals, Managers and Specialists (IPMS) – the trade union of health and safety inspectors – also agrees. It takes the view that the number of agricultural inspectors in the field (at the time of writing, 142) is completely inadequate to cover the 300,000 premises and 700,000-strong workforce.[3] Even if each workplace was visited on a rotational basis, an inspector would still only call every 9.8 years. IPMS calls for the number of agricultural inspectors to be increased by at least 100.

HASAWA and COSHH

HASAWA (Health and Safety at Work Act 1974) puts general duties on employers to safeguard the health of workers and members of the public, 'as far as is reasonably practicable'. Employers have a duty to train, inform, instruct and supervise workers about health and safety. Duties are also imposed on manufacturers, suppliers and importers to research and test substances. Under HASAWA, the Safety Representative Regulations enable trade union representatives to receive information and to investigate chemical substances such as pesticides. They may be able to improve standards for their members and protect the public too.

Placing the responsibility on the employer

COSHH (Control of Substances Hazardous to Health Regulations 1988) replaced the old legal controls in the Poisonous Substances in Agriculture Regulations 1984. Employers must now assess the risks of all chemicals, including pesticides. The Regulations lay down the essential requirements and a sensible, step-by-step approach for the control of substances such as pesticides, and for the protection of people exposed to them. Having assessed the possible risks, the employer has to decide which control measures are appropriate. The emphasis is on preventing exposure. If this is not reasonably practicable, adequate controls such as enclosure, ventilation or safe systems (where the employer lays down a strict procedure) have to be introduced.

The COSHH approach differs from FEPA and previous laws dealing with pesticides – it is based on good occupational hygiene practice. In the past, control of exposure was exerted through the label and through wearing protective clothing, and was therefore dependent on the operator following instructions. Now the onus is on the employer to prevent the operator being exposed to the chemical.

Enforcement

COSHH and COPR have been used to prosecute a variety of incidents. Wolverhampton Council were fined £1,250 for allowing a 19-year-old apprentice carpenter to spray, instead of brush, a fungicide on the ceiling of a council property.[4] A self-employed contractor in Berwick pleaded guilty to not taking 'reasonable precautions to protect human beings, creatures and plants in the use of the pesticide Triazophos. Despite warnings on the label, he had sprayed a rape crop in flower and had not warned local beekeepers. A large number of bees were killed.[5] And in November 1990 a farmer near Preston pleaded guilty to using a pesticide incompetently. HSE inspectors made a spot check and found him spraying in winds and using a double dose. His defending counsel said: 'He hasn't been on courses or anything, but there is a lot to be said for 26 years' experience.'[6]

Codes of practice

With these different laws there can be some confusion, and so to clarify matters a code of practice has been drawn up. *Pesticides: For the Safe Use of Pesticides on Farms and Holdings* incorporates both FEPA and HASAWA, and has a similar status to the Highway Code. You would not be prosecuted for breaking it, but if you had a accident or were prosecuted under COSHH or COPR it would be quoted as evidence in court.

The code of practice deals with the following aspects of the use of pesticides:

- duties of users
- training requirements
- transportation
- planning the use of a pesticide
- preventing exposure
- application methods
- using control measures
- protection of the public
- monitoring exposure

- health surveys
- disposal of leftover pesticide and containers
- keeping records.

Another similar code, *For the Safe Use of Non-Agricultural Pesticides*, produced by the Health and Safety Executive, covers pesticides used in timber treatments, for public health purposes, in stored food products, amenity horticulture and commercial forestry. There is also a code called *Control of Substances Hazardous to Health in Fumigation Operations*.

Disposal

Laws cover not only how pesticides are introduced and used, but also where they can go. In addition to FEPA and HASAWA, the Control of Pollution Act 1974 (COPA), the Water Act 1989 (WA) and their respective regulations apply to the disposal of pesticides and their containers. Under COPA it is an offence to abandon or dispose of poisonous, noxious or polluting waste, including waste from agriculture, on any land where it is likely to give rise to a hazard. Under the Water Act you cannot allow poisons to enter legally controlled waters. Discharges into water or sewers of pesticides in the Red List (see Table 3) require approval from the Secretary of State for the Environment before consents can be given by the National Rivers Authority or your local water company.

If you are in any doubt, contact the Environmental Health department of your local authority. The staff may give you advice themselves, or refer to the waste disposal authority dealing with your area. Don't be tempted to empty unused containers into any water system; that includes washbasins, toilets and drains. Presume the containers are dangerous and take advice. Don't pretend it has nothing to do with you.

Residues

Residues in food leaving the farm gate are, as explained in Chapter 10, controlled by the Pesticides (Maximum Residue Levels in Foods) Regulations 1988. These Regulations set levels for about 50 pesticides, in accordance with EC directives. Food which exceeds MRLs can be seized and destroyed.

The Food Safety Act 1990

The new Food Safety Act 1990 gives the Environmental Health Officers (EHOs) of local councils tougher powers over food contamination. In particular, there are a number of powers in this law which can be made to apply to pesticides. One is particularly important: it is the so-called 'due diligence' section Act (Section 21). This states that makers or sellers of food have to do everything in their power to ensure that the food is not 'injurious to health'. So if the food is not all right, the maker or seller should be in trouble, unless they can show that they showed 'due diligence' or care to do everything in their power to prevent contamination. The Act also says that the consumer has the right to food which is 'of the nature of substance or quality demanded' (Section 14). The law says you should get what you want (demand). Finally, the Act gives EHOs one other vital new power. It states that long-term ill-effects should be included in any judgment of a food complaint (Section 7.2). The Food Safety Act 1990 replaced large chunks of the Food Act 1984.

Other legislation that covers pesticides

Various pesticide issues are covered by other laws. Relevant ones are the

● Medicines Act 1968
● Agriculture Act 1970
● Food Act 1984
● Food and Drugs Act (Scotland) 1956
● Cosmetics Products (Safety) Regulations 1984.

Some pesticides are covered by the

● Poisons Act 1972
● Poisons List Order 1982.

Keep abreast of changes

The laws on pesticides are continually changing, and many more general laws have a bearing on pesticides. For the latest information, telephone the local office of the Health and Safety Executive; the Ministry of Agriculture, Fisheries and Food; or your council Environmental Health Department. They should be able to help.

How Can You Complain, and What Response Can You Expect?

•

Reporting incidents of pesticide poisoning

Y ou may believe that you have symptoms of pesticide poisoning from exposure at home, in the garden, in a park, from
spraydrift in the countryside, or at work.

- Find out the common name of the pesticide – use Chapter 4 to help
- Look up the name in Table 2 or Table 3. Table 2 will tell you what group it belongs to
- Find the relevant group in Table 4 to see if your symptoms correspond. Be wary about jumping to conclusions
- If you believe you have been affected. . .

Go and see your doctor. Take with you a list of your symptoms and the relevant list from this book. Here are the routes to
follow if you want to complain. We have ordered this chapter
according to where your problem started.

In general

- whoever you complain to – whether doctor or Health and Safety Inspector or Environmental Health Officer – make sure your complaint is passed on to the Pesticides Incidents Appraisal Panel (PIAP). Whatever the complaint, it must be fed to the PIAP. How PIAP works is outlined on p. 123
- keep a record of what happened. Note where it was, and what was involved. Ideally keep a sample, if that is possible
- take a note of any relevant names – the name of the pesti-cide, who sprayed it? Where are they based?

At home

- if you have been affected by pesticides from wood treatments or food, contact the Environmental Health Department at your local council offices. The telephone number will be in your local phone book. Environmental Health Officers and Trading Standards Officers are the people whose job it is to ensure that your food is safe and fit to eat, and that when you buy food you get what you actually want

- if the Environmental Health Officer needs to be encouraged, for they are under-resourced, say you want the complaint investigated under the Food Safety Act 1990. The Food Safety Act is summarised at the end of 'What laws govern pesticides use?' (Chapter 13). So you could encourage your local Environmental Health Officers or Trading Standards Officers by asking them to check and toughen up standards for pesticide residues
- if you think you were affected by food, keep a sample. This is vital evidence. In every area of the UK there is a public analyst. Public analysts use independent scientific laboratories and can test the food, find out what residues are in it, and make reports to the council or a court of law, if your complaint gets to that stage.

In the park

- contact your local councillor to register your complaint
- find out whether local authority workers or outside contractors are responsible
- find out if the local council has a policy on pesticide use
- ask for the Environmental Health Officers to investigate
- if you work in a public park and use pesticides, register your complaint with PIAP.

In the countryside

- contact the local agricultural inspectors directly. They are in the phone book under 'Health and Safety Executive – HM Agricultural Inspectorate'
- ask for your complaint to be formally investigated
- if you are working with pesticides, register your complaint with PIAP.

At work

- make sure it is recorded in the accident book

- ask your doctor to pass your complaint to the Employment Medical Advisory Service (EMAS). This is a confidential medical service specialising in ill health at work, and will not reveal anything to your employer without your express permission. EMAS may investigate the problem directly, or it may pass the complaint on to the local HSE inspectors
- you may wish to contact the HSE inspectors directly – their number will be in the phone book under 'Health and Safety Executive'
- you can also contact your trade union safety representative, who has a legal right to receive information from the HSE
- the trade union safety representative can also get more information from their union. Some unions have specialist health and safety officers. All have advice and back-up. Three unions have been involved with surveys of workers for pesticide exposure. They are the Transport and General Workers' Union (TGWU), the General, Municipal and Boilermakers', and Allied Trades Union (GMBATU), and the National Union of Public Employees (NUPE).
- register your complaint with PIAP.

What is PIAP?

PIAP, the Pesticides Incidents Appraisal Panel, has representatives from the Employment Medical Advisory Service (EMAS), HM Agricultural Inspectorate, the Department of Health's Toxicology Section and the National Poisons Information Service. It investigates all incidents where the use of a pesticide during a work activity may have affected your health. PIAP also consider cases involving veterinary medicines which contain pesticides – e.g. sheep dips.

PIAP receives information about the incident in a report from the relevant enforcing inspector, and any investigations carried out by EMAS. Each case is considered in relation to the known effects of the reported pesticides. PIAP classifies the incident as

- *confirmed* – the pesticide caused the reported adverse effects
- *likely* – poisoning suggested but not proven
- *unlikely* – little likelihood that the pesticide was responsible, but it cannot be ruled out

- *not confirmed* – no evidence that the pesticide caused adverse effects
- *insufficient data* – not enough data to make a proper consideration.

Make sure you don't end up with the last category. Get the relevant inspectors to investigate as soon as possible – while the evidence is still available.

The results of the PIAP investigations are published each year.[1] These are the official figures which we quote elsewhere in this book. We are sure that there are many more complaints that are not investigated as people frequently do not know where to go to complain. Let's change that and make the official statistics reflect what really goes on.

Complaining about pesticides in principle

So much for the practical aspects of complaining about a pesticide poisoning incident that has affected you. But what if you want to register your disapproval of using pesticides in the first place? Over 20 years ago J.M. Barnes of the Toxicology Research Unit at the Medical Research Council Laboratories at Carshalton in Surrey made a famous, categorical statement: 'There seems not a tittle of scientific evidence that pesticide residues in human food present a real threat to health.'[2] Ever since then the scientific community has quietly and completely split. There are arguments and evidence on all sides – this book is a testament to that fact. But, as we have said a number of times already, we think *you* have the right to be included in the debate about pesticides. What happens if you actually join in the arguments – as a mum or dad, as a worker who uses pesticides, or a gardener or grower? The rest of this chapter gives you an early warning of the kind of arguments you will get back if you merely question pesticides, let alone if you make it clear that you want less or no pesticide use.

The old arguments

In the 1950s pesticides were offered as the scientific, sensible tool for the modern farmer. People who were against pesticides were few. They were dismissed as the 'muck and magic' brigade. Agrochemical companies did not have to take their critics very seriously. Modern farming, the companies argued, could produce more, with fewer workers. The new food system would lower the price of production and free people to work elsewhere.

This happened. And not just for farm workers. Fifty years ago, there were around half a million farmers. In 1960 their numbers had dropped to around 350,000. Today there are about 150,000. That means fewer people to look after the land. Pesticides are the liquid hoe, and the liquid labour replacer. David Naish, deputy president of the National Farmers' Union, put this old argument clearly in 1990: 'Pesticides keep the cost of food down because they replace labour. In 1938 there were one million employed on the land, producing one-third of the food needs of 48 million. By 1988 there were less than half a million full-time farmers and workers producing three-quarters of the indigenous food needs of 57 million.'[3]

A quiet revolution in the making?

Today, the industry is more on the defensive. 'Muck and magic' is no longer a term of abuse but the title of a highly successful TV organic gardening series, hosted by people from the excellent Henry Doubleday Research Association which runs the organic trial gardens in Ryton, near Coventry (well worth a visit, by the way – whether you are for or against pesticides). And farmers are increasingly interested in going organic. Mr John Selwyn Gummer MP, the Minister of Agriculture, has introduced some financial support for non-agrochemical farming. (No need to get excited – it amounts to far less than 1% of total Ministry research support!)

So where does this leave the pesticide industry? Doomed to a steady decline? Hardly. It wants to survive, and is determined to face its new circumstances. This new context for agrochemical-dependent farming includes:

- public concern about the environment
- polls suggesting people want more organic, pesticide-free food and/or labelling which tells whether a pesticide has been used[4]
- a loss of consumer confidence in the Ministry of Agriculture, Fisheries and Food.

As a result, the companies which manufacture pesticides and those that sell affected products are rushing to their public relations companies. In the 1989–90 financial squeeze PR business boomed, even as advertising business felt the pinch. Already new, smoother approaches are emerging from the agrochemical and food industries.

What the pro-pesticide lobby will argue

Here are a number of the reactions and arguments that you may meet if you express concern about pesticides:

● *'Pesticide use is based on sound science'*

The first response from the pesticide companies is often: 'How dare you question our scientists!' People in white coats will get wheeled out to say, 'Don't worry, it's all quite safe.' One review of this type, published in 1986, went so far as to say that 'speculations that intensive farm production has undesirable effects on public health are not based on fact'.[5]

The truth of the matter is that there is a debate about intensification of agriculture – both in and outside science. For instance, in the United States a report from the National Research Council estimated that beef was the second most risky food from pesticides.[6]

When pushed, the argument often takes a peculiarly British turn. You get accused of Luddism. Named after a mythical figure, Ned Ludd, 'luddite' was a term used for people in the early 1800s who broke up the then modern agricultural machinery because it threatened their jobs and way of life. The word came to mean being anti-progress. Serious stuff, but is it anti-progress to question pesticide health and safety? The issue is not whether you or we are for or against progress, but whose definition of progress and on whose terms progress is to be. It is perfectly possible to rear a beef animal without feeding it pesticide residues.

● *'without pesticides, the world will starve'*

This argument is discussed in more detail in Chapter 15. Only three points need to be made here.

First, there is a lot of starvation *with* pesticides. UNICEF calculates that 14.6 million children die every year from malnutrition. In 1980, according to the World Bank, 730 million people had not received enough calories for an active working life.[7] The reasons for this dreadful situation are much more complex than just whether pesticides are used or not. As Christian Aid says: 'The basic causes of hunger are economic'.[8] One of the best ways of tackling hunger is to support women, who already produce around 50% of the world's food, yet own only 1% of the world's land.[9] Around 40% of the world's grain production is fed to animals, so alarm at consumption of grain exceeding production in the late 1980s should be tempered by realising that feeding animals less wastefully has to be part of the solution to hunger. Be wary of those who argue simply that there are too many people and too little food.

The second point is that the reason people starve has less to do with pesticide use and more to do with distribution and money. In general people eat well, badly or not at all according to whether they have money, access and information about food. In poor countries – or for the poor of rich countries, even – money dictates whether people can buy food, and what they get. That is why in the middle of the infamous Irish potato famine in the last century, food was actually being exported! Many reviews of hunger in this century have come to similar conclusions.[10]

The third point is that, however good pesticides are at killing pests, the boomerang has a habit of coming back. Pesticide use is often followed by the pests becoming resistant. That is why the newspapers carry stories about 'superbugs' or 'super rats'.[11] Environmental Health Officers, like doctors with antibiotics, sometimes find that their pesticides are no longer working so well. The result is a treadmill. New pesticides have to be developed to combat the resistant pest. And so on and so on. US entomologist Professor Robert van den Bosch summarised the treadmill argument in the case of DDT: '. . . only a third of a century after the discovery of DDT's insect-killing powers, and despite the subsequent development of scores of potent poisons, the bugs are doing better than ever, and much of insect control is a shambles.'[12] So should we jump off the treadmill? Sometimes it seems unthinkable. One researcher, Professor David Pimental, has calculated that the removal of all pesticides from US agriculture would raise crop losses from pests from the current 33% to 42%.[13]

- *'natural toxins are a bigger danger than pesticides – it's all a question of relative risk'*

This argument goes like this. Natural toxins in fruit and vegetables are more important poisons than any pesticide residue. This argument has been put most strongly by Professor Ames of the USA, well-known for the Ames test (see Chapter 6), who argues that public concern has got dangers from pesticides out of proportion. The American Consumers' Union has replied at length to this argument, along the following lines:

- saying that natural dangers exist is no excuse for adding to them
- current scientific evidence does not fully justify Ames's case
- the benefit of the doubt should go to the consumer anyway.[14]

Chris Major, former Director of the British Agrochemical Association and now ICI Agrochemical's public affairs manager, has said: 'Today we will ingest 10,000 times more naturally occurring pesticides in what we eat and drink than those made by the chemical industry.'[15] In the potato alone there are nearly 150 naturally occurring toxins. And for the average person in the UK, who consumes around 120lbs of spuds in a year, this means eating just over 3oz of solanin. (Solanin is found particularly in green potatoes and is particularly undesirable for pregnant women.) Mr Major then repeated one of Professor Ames's statements, that 99.9% of all toxins that humans take in are natural.

The US Consumers' Union took Professor Ames to task for his approach. In a detailed criticism it said that he 'put too much emphasis on catchy phrases like "99.9 percent all natural", which are not scientific facts, but rather your own intuitive assumptions, opinions, or crude calculations'.[16]

So what is our answer to this 'natural' argument? No one will dispute that there are natural toxins. The point is that pesticides are a controllable and avoidable extra burden to our bodies. People have little control over natural toxins, but can control – and expect controls on – artificial pesticides. One theory says that the human body has had thousands of years to develop ways of coping with natural toxins, but that pesticides have come relatively recently. Take deadly nightshade. It can kill humans, but through the ages people have learned not to eat it. Humans have survived precisely because they learn and adapt. Pesticides are an avoidable risk.

● *'Our standards are good: a pity about the foreigners'*

This argument is much used in Britain or USA or Australia or wherever you happen to be arguing. It implies, a bit smugly, roll on the day when the others catch us up. We think there is little cause for national smugness (see Chapters 3 and 5, and Table 3). The UK figures on pesticiedes residues do indicate worse levels for some imported foods, true. The 1989 government report, for instance, implied that there were problems only in imported foods. Particular emphasis was given to meat imported from China. The 1990 report from the WPPR did the same.[17]

Less emphasis was given in the report to worrying results about baby food. Of 107 samples of infant food tested in 1988–9, residues were found in 22. Of 266 cereal and cereal products, 15 (6%) were over the Maximum Residue Limit. Forty (6%) of 667 fruit and vegetable samples were over the MRL too.[18] We are far from reassured.

In 1989 one short review of international standards by the London Food Commission found 14 pesticides in use in Britain which other countries ban:

● Amitrole, a herbicide, is banned in Sweden, Finland and Norway
● Carbaryl, an insecticide, is banned in Germany
● Paraquat, a herbicide commonly used in gardens, is banned in Sweden, Finland and Germany
● Phenylmercury acetate, a fungicide used on seeds, is banned in Turkey and New Zealand.[19]

See Table 2 for pesticides used in UK and their bans elsewhere.

From the public's point of view, nationalism should have little place in setting scientific standards. The goal should be the best standards everywhere. In principle, standards should rise to the toughest applied anywhere, but this will not happen automatically, particularly when there is pressure to remove barriers to trade in Europe by 1992 and under GATT by 1993. (see Chapter 11).

● *'Our detection methods are so good these days'*

This is a lovely one. It says that the public is being misled by figures of residue detection into thinking that pollution and adulteration are widespread. Equipment these days is so sophisticated, the argument runs, that it can pick up unimaginably small residues. There is nothing to worry about.

The British Agrochemical Association is sensitive about residues in water. It takes great exception to the European directive on water quality, which sets so-called Maximum Admissible Concentration limits (MACs).

● for any one substance, the Maximum Admissible Concentration limit is 0.1 part per billion (ppb)
● for total pesticides, the Maximum Admissible Concentration limit is 0.5 parts per billion.

The agrochemical companies want this changed. '0.1 ppb is equivalent to the thickness of a credit card in the entire distance between London and Peru' it says, adding:

Ten years ago it was impossible to measure organochlorine compounds below 0.1 ppb. But now the sensitivity of analytical techniques has increased more than one hundred fold and it is possible to measure these compounds down to 0.001 ppb. *The result is that chemicals which were thought not to be present in water are now being found in measurable quantities. This does not however necessarily mean that these quantities have any significance for environmental or human health.* [Our emphasis.][20]

What is our reply to this criticism? Simply that the better measurement now shows how our bodies and the environment are subjected to a persistent, slow drip-feed of residues. Instead of concern about pesticides being dismissed as hot air, measurement now shows that it is indeed happening.

Agrochemical companies try to reassure you about how good their scientific measurement now is. The credit card metaphor is almost boast. But human smell is also wonderfully sensitive. One part per billion is roughly the threshold for humans to detect the scent of garlic. Guinea pigs may die when exposed to 1 ppb of TCDD Dioxin, a contaminant of certain pesticides. So even tiny amounts can be critical to humans. Small can be significant. We think that the one credit card metaphor is not just misleading, but could even be found to be wrong in the future. Take the contamination of water. Some commentators argue that levels of pesticide are too small to worry about. The BMA report was less assured and expressed concern about drinking water. Another concern is not just that residues are there, but that levels in some cases are rising.[21] The insecticide Lindane is now found in the North Sea at a level that has doubled in the last four years.[22]

Pesticides are tested on animals, but it is unethical to test them on humans. Unwittingly, people are therefore guinea pigs, if pesticides are approved and then monitored in use (see Chapter 6 on animal testing). No one knows what the long-term effects of low doses of chemicals are on the human body.[23]

● *'Don't listen to the message – shoot the messenger'*

This is an easy tactic, but it often backfires. And it is used less and less when the arguments get hot. The industry has learned that using this argument may misjudge the mood of the public, in these Green and publicity-minded days when PR people front for and advise agrochemical and food companies, and governments (see Chapter 14). Shooting the messenger is a sure sign of an industry or profession which has not had to justify itself too often.

Look what happened when the London Food Commission's report on pesticide residues was published in 1986. It was written by Peter Snell, a trained food technologist with a degree in food science and another in economics; he was also a paid-up member of the relevant professional body, the Institute of Food Science and Technology. On BBC TV one academic nutritionist was asked to comment on the LFC report. He made little comment about its content, but chose instead to criticise the messenger: 'They have been termed by a colleague as food terrorists. That is, they hold the Government to ransom, by making all sorts of wild claims. . . . Then they force the Government into taking some sort of action on the basis of a bandwagon-type lobby.'[24] The fact that the colleague referred to was a professor who did consultancy work for the food industry, and the fact that the LFC report had merely reviewed the existing scientific literature to report on the problems found for each pesticide, was apparently judged less important than the threat to the status quo. The LFC report was asking for tougher controls, not banning, note. In fact the report's subtitle was 'the care for control'.

Another version of the 'shoot the messenger' syndrome is to

accuse the messenger of being 'emotional', as happened in the Daminozide story. Parents for Safe Food was set up to call, not for its banning, but for its suspension. This was made quite clear in the letter signed and presented to the Prime Minister at 10 Downing Street by the celebrity founders in May 1989. The line was 'suspend, until doubts have been removed' and give 'the benefit of the doubt to the consumer'. Reasonable stuff. Initially neither industry, nor government, nor advisers seemed quite sure how to treat Parents for Safe Food. A seasoned voice, Professor David Conning, formerly of the British Industrial Biological Research Association (BIBRA), called PSF 'neurotic whingers,' in 1990. Hardly objective science!

At a meeting of Friends of the Earth in 1989, to discuss the use of chemicals in food production, Mr Chris Major of ICI saw the arguments against use of chemicals as basically an 'emotional issue'. He accused groups like FOE of issuing 'half truths'.[25]

● *'Talk down the opposition'*

At its worst, this tactic can mean pulling out all the stops to wipe you off the floor. In his book *The Pesticide Conspiracy*, US academic Professor Robert van den Bosch spills the beans on many pesticides issues, One story is relevant here. When asked to prepare an article for *Life* magazine, Professor van den Bosch wrote about exciting non-pesticides methods of controlling pests. The article was first delayed, then typeset and mocked up. He had a meeting with the editor and commissioning editor, at which the editor said, '. . . there *was* pressure from the chemical industry to kill the article'. (His emphasis.) The article never appeared.[26]

These days, happily, such crude tactics concerning pesticides are rarer. But it does still happen over other issues in the world of agrochemical farming. One academic researching the controversial milk hormone Bovine Somatotrophin (BST) – now banned, happily – has claimed his experimental herd was dispersed and his contract ended while he was out of the USA on a lecture tour. He alleged he was getting worrying results from his trials.[27]

Rubbishing the opposition still goes on, but in different forms. Pesticide companies are a little worried about the public's interest in organic produce. The companies spend a lot of time these days rubbishing organics as cranky or as a sure recipe for worm-eaten carrots. At a symposium on pesticides in the 1990s, ICI's Chris Major pulled out his crystal ball and said: 'The early 1990s will be characterised by consumer disillusionment with the quality and reliability of organic produce. It will remain a niche market, but the heady forecasts of the late 1980s that 20% of food could be grown organically by the year 2000 will not be achieved.' He said supermarkets would drop selling organic produce. One did in 1990, but as a general trend, it seems unlikely according to business analysts.[28] Whether it is Prince Charles's organic Highgrove bread for Tesco's or organic wine sales, the market prospects for pesticide-free foods are good. Opinion polls suggest strong support for growing food without pesticides. ICI quote one that concludes: '4 out of 5 people believe that we can do without the use of pesticides.'[29]

Increasingly, pesticide companies are trying to improve their public image. Here is a simple illustration. In May 1990, Maggie Yarwood of ICI Agrochemicals was given a two-year secondment to work with the British Agrochemicals Association to establish a regional public relations approach. Her task was to set up a regional network of people who work for agrochemical companies. Her 'target'? 'The industry's natural allies such as farmers, agricultural students, merchants and rural interest groups', as well as councillors, doctors and council executives.[30]

- *'Let's meet and talk'*

This is industry at its most civilised and reasonable. We have ourselves been happy to meet and engage in dialogue about pesticides.

In August 1989, what was described as an unlikely alliance of consumer, trade union and environmental groups, concerned about pesticides, joined forces with the British Agrochemical Association to launch an attack on delays in the pesticide registration system. This came after years of private and informal meetings, begun in 1987, to see where there could be common ground.[31] The group called for

- completion of safety reviews to be speeded up
- more inspectors to be appointed
- more frequent testing for pesticide residues
- and a new National Pesticide Incident Monitoring Scheme.

On 31 May 1990, ten European groups published the European Ecological Consumers' Pesticide Charter.[32] Both the National Farmers' Union and the British Agrochemical Association were quick to welcome the Charter and to ask for talks. So did the National Farmers' Union. We welcome such dialogue.

It can be useful, tactically as well as to safeguard public health, to join forces to lever better protection measures from government. There is a danger, however, in getting bogged down in talks, which stop you working on your central task. So be warned!

● *'We are Greener than thou'*

This most amusing tactic is relatively new. In the 1980s, the chemical companies began to realise that there was wisdom in stealing the opposition's clothes. These days it is hard to pass an advertising board or watch the TV or read the annual company report without seeing a chemical company claim that it is as Green as could be.

Environmental and food groups are campaigning hard to stop misleading claims. The new Food Safety Act 1990 (Section 15) supposedly stops misleading claims. No such powers yet exist to be tested in court over claims of 'Greenness' but a new eco-labelling scheme is to be introduced by the end of 1991, either by the UK government or the European Commission (EC). Experience from Germany's Blue Angel labelling scheme suggests consumers should be wary. The EC or like schemes should be an improvement, though much depends upon whether independent voices are on key monitoring or approved committees.[33]

Most pesticide advertisements proclaim their 'macho' kill-factor. A Green pesticide image would make an interesting but not impossible challenge for advertisers! Undoubtedly it will come. Already businesses are keen to do environmental audits of themselves, on terms with which many environmental groups are less than happy. It is easy to sell the line: out of sight, out of mind, like the famous petrochemical advertisement which showed beautiful countryside, but not the factory over the hill.[34]

Can the World Be Fed Without Pesticides?

In a word, yes. But of course not everyone is prepared to admit this. In the first paragraph of its most recent report the British Agrochemicals Association said: 'Many people, influenced by the media, view pesticide companies more as polluters than as vital protectors of the food supply.'[1] Do they really?

Pesticides make pest control easier. They reduce labour requirements, especially skilled labour. They reduce crop losses, but not completely. No pesticide gets rid of all the target pests. The very presence of so many chemicals shows their singular lack of success at getting rid of all pests. But none of this has much to do with world food production. The main driving force to make pesticides is to make profits. That may not be terrible. But the companies should be honest and not pretend to be the bastions of conscientiousness. Consider the following:

- the chief of the Plant Protection Service of the FAO (Food and Agriculture Organisation of the United Nations) estimated in the mid-seventies that 800 million lb of pesticides were used a year in under-developed countries. The 'vast majority', however, were for export crops, principally cotton and to a lesser extent 'fruit and vegetables grown under plantation conditions for export'[2]

- much more food could be produced. The best land in many countries is used for export crops, including cotton and rubber. At the 1974 World Food Conference it was estimated that world food could support a population four times the present. This may be achieved by changes to the following two items[3]

- water distribution. Water is the most important nutrient of all, and its redistribution would transform world agriculture. After irrigation had been introduced, the deserts of Peru produced crops after two thousand years of dereliction. Major increases occurred in Mexico in the 1960s following

irrigation. The Sahara sits on the largest underground sea.
While the potential is enormous, such changes need not all
be on a grand scale. Small earth dams, sprinkling, controlling evaporation, and recondensing sea water for example.[4]

● land ownership. The FAO has estimated that only 60% of
the land available for crop production throughout the world
is actually used. 17 million acres of the UK (about ⅓ of the
total surface) is left to rough grazing and grouse shooting,
More food could be produced in this country. Look at what
happened during the two world wars.[5] What is at issue is the
impact of intensification on the environment, and whether
current farming methods are desirable or sustainable.[6]

The concentration of agricultural production on certain land
intensifies the use of machines and chemical fertilisers. The
problem of pests is also increased, encouraging the use of pesticides. The drive is for labour efficiency on the best land. In
Europe, a triangle of intensive, monocultural production has
emerged running from East Anglia up to Denmark and down
to the Paris Basin. Around 80% of the European Community
farm budget goes to 20% of Europe's farmers in this triangle.
Production rises by around 3% per year. As a result a 100
hectare cereals farmer probably needs to take one-fifth of this
land out of production to ease the financial burden to Europe's
tax payers of paying for over-production.[7]

The 'problem' for the controllers of world food supplies is
not how to produce more food – but how to control and often
contain production. For some commodities, this can mean having to produce less, and still make a profit. Look at what has
happened to US and European agriculture in the 1970s and
1980s. The saga is one of over-production, resulting in surpluses! Whenever a lot of a particular food is produced, its price
goes down – witness potatoes in the summer of 1990 in the UK.
Price plummeted and kept inflation below 10%![8] It is what is
called 'the market mechanism'. Food producers do not like that.
Small producers cannot do much about it. But larger companies
can, as they can influence the policy of any country.

The history of farming in many developed countries is the
history of controlling production and overcoming 'the
market'. Look at the guaranteed prices with quotas introduced
to support UK farming in the 1930s. Witness the set aside policy in operation in the UK at present: farmers are paid £80 an
acre *not* to produce anything, and 2.3% of all available land in

the UK is 'set aside'! Then there are the milk quotas introduced by the EC in the late 1980s. All these deal with *excess* production. We have all heard of butter mountains and wine lakes. Two less well-known solutions are the transformation of food into something else, such as drugs, or petrol or fuel, and the massive export subsidies offered to get rid of the stuff.

Growing food by relying upon pesticides is just one among several approaches. Often the doomsday views which argue that there is a crisis of too many people, too little land and not enough food assume that there is only one way to grow food. There isn't. Because companies have carved out huge and profitable sales markets for agrochemicals, including pesticides, pesticide-dependent agriculture has had vastly more investment than other systems. There are systems which grow food without pesticides at all – organic, permaculture, biodynamic (see Chapter 16).

There is a simmering trade war between the USA and Europe. Both claim that the other is using 'unfair subsidies' to bolster its farm production. The USA is concerned at not being able to export into Europe and at the way the EC 'dumps' on the rest of the world. The EC is concerned to maintain its own farming and not be dependent on imports. At the 1990 summit it was resolved that each would investigate standards and criteria for subsidies and that there would be regulation of phytotoxins – clearly linking the 'problem' of overproduction with the use of herbicides. (See the section on GATT in Chapter 11) People starve in the world not because the food cannot be produced but because they cannot afford to buy it, move it or because they come up against various other political barriers.[9]

One of the authors of this book embarked on a career in agricultural science in the belief that his scientific endeavours could help save the world from starvation. It soon became obvious that only certain groups could afford the pesticides, and it was they who determined the direction that agriculture took. White-coated scientists walking the world do not help starving people get out of the grip of foreign technology and debt. Well meaning science needs to understand global power politics. And it is that which determines who gets fed.

Some would have you believe otherwise. According to an annual report from Bayer, the third largest pesticide manufacturer in the world, 'In view of the challenge posed by world hunger, emotional attacks against conscientious agricultural chemicals research are attacks against humanity.'[10]

What Are the Alternatives?

One of the biggest questions you or we ever get asked is this. We know that pesticides are a problem, but if you had to choose between using them and starvation, what would you do? The short answer is that we would probably use them, but as little as possible. Alternatives to the use of pesticides are essential. They are also, as we have shown earlier, under-researched and under-supported. The 1990 budget for organic research and development by the UK government was £1 million. This was a grand 0.33% of all the public money spent on agricultural research![1]

Fear not. Many organisations particularly from environmental and trade union movements have been dedicated to reducing or eliminating pesticide use. They have grown and have won public support. These organisations include Friends of the Earth, the Soil Association, Greenpeace International, the Henry Doubleday Research Association, the Women's Environmental Network, the Pesticides Trust and Pesticide Action Network and many more. The number of producers who are questioning and leaving agrochemical farming is growing in the UK. Supermarkets and food processing companies throughout Europe and the world are seeing demand for pesticide-free products grow apace. Even conservative trade estimates aim for 5% of the food market being organic by the year 2000.

This chapter, addressed to everyone who uses pesticides, at home, in the garden, at work, on the farm, indeed everywhere, looks at alternatives to pesticide use, from A to Z. Our general advice is: don't be put off looking for alternatives. Often success is a question of equipping yourself with different skills. Sometimes these are new skills, sometimes old ones. Sometimes you won't find what you want. Most people admit defeat over slugs and lettuces.[2] Any effective answers, please write to the HDRA.

- *Accept a few pests.* Nothing, but nothing, gets rid of all pests. A pest or its damage are not usually tolerated, but the unseen pesticide residues are. Get rid of the 'agro-macho-techno-fascination' attitude that humans are always 'at war' with the bugs.

- *Biological control* means control by natural enemies. It can be cheap, effective and permanent. However, it may disrupt other parts of the eco-system. A biological control can get out of control. Having eaten the first pest, it may die out or start affecting something else. Biological control, ideally nonetheless, should be the bastion of pest control. There are three prongs.

First, consider the use of existing systems. In California each year 7.5 billion ladybirds eat 3.75 trillion greenfly in alfalfa crops. We don't know who counted them all, but we do know that without the ladybirds there would be trouble!

Secondly, the food and habitat of natural enemies, including birds, cats, lizards and microbes, can be encouraged.

Thirdly, classical biological control involves the deliberate importing of natural enemies. Because many pests are of exotic origin, the best place to go is the country of the pest's origin, from which to bring back some local predators or parasites. Well over 200 such successful examples exist. It is not new. The most famous case centred around getting rid of the prickly pear cactus in Australia, where in the mid-1920s it infested 60 million acres. A combination of insects that fed on different parts of the cactus was introduced. The most important was a moth borer called *Cactoblastus* sent from Argentina. In one area it destroyed 90% of a 100-mile belt of the cactus within two years. A later example is from Barbados in the 1960s, where parasites were used to control the citrus fly. The tiny *Trichogramma* wasp has been mass released and controls moth pests on million of acres throughout the world. There have been some cases where the introduction of a foreign predator has caused problems too.[3,4]

- *Calculate the costs.* Find out the right pesticides. Get the proper equipment. Investigate the possible damage to the environment, possible legal responsibilities and possibility of illness. And work out whether using a pesticide is even worth it.
- *Develop fungi instead of weedkillers.* Fungi can destroy crops. They can also destroy weeds. Such fungi, called mycoherbicides, promise to deliver what other herbicides cannot: a

highly specific treatment aimed at a single species of weed that will leave other plants untouched. Such fungi are chosen from naturally occurring fungi, and are not biotechnological creations. They have to be produced on an industrial scale. In the 1980s fungal diseases were marketed in the USA to control weeds in soyabean and citrus crops. The greatest success so far are when used in combination with a chemical, by helping to reduce the amount of chemical for the same effect.[5]

● *Examine the label* and get hold of the data sheets, if you are using a pesticide. Campaign for the right for consumers to have a P for Pesticides label on foods. In a government survey, half said the label should say what pesticide has been used and a quarter wanted the name of the pesticide used.[6] The right to examine the label should be won!

● *Farmers* could collect set-aside money (money being paid for not growing a crop) for five years, by which time they would qualify for going organic. Talk to British Organic Farmers and Organic Farmers and Growers (see addresses under 'Soil Association' on p. 234) for advice on the most difficult phase: transition from treadmill farming to organic farming.

● *Greenfly* don't get on with soapy water. Or squirt them with a strong jet of water. They fall off and cannot get back.

● *Hygiene* probably does more than anything. Cleaning up thoroughly in the home gets rid of the debris that forms the food of many pests – flies, ants, cockroaches, mice etc. Occupational hygiene is the basis for reducing exposures to pesticides and for the 1988 COSHH Regulations. And field hygiene – keeping crops clean and clearing up residues quickly – keep pest habitats to a minimum.

Read up on domestic hygiene. Ask your local council's adult education officer to put on an evening class. Make sure you demand better hygiene from public places. Hygiene is not your private problem. It is a public health affair.

● *Integrated pest management/control (IPM)* is based on an understanding of the pest problem in the full environmental context. It requires knowledge, skill and labour. It is an ecological approach – one that tries to find out the natural history of the pest and then aim at particularly vulnerable parts using methods that are not likely to damage anything else (see *biological control*, which is one form). Cultural controls based on 'good traditional methods' are another.

Examples include removal of plant residues by ploughing, flooding, long leys (long-term grassland) and hoeing; interplanting crops with beneficial covers (plants being used to cut down weeds to allow your crop to flourish), pest deterrents and trap crops; planting a variety of crops; crop rotation; improved drainage. Physical controls can involve blocking or trapping insects (see *traps*), controlling temperatures in greenhouses, light traps, screens, metal barriers to stop rodents, hand picking of insects etc.

IPM also includes the more judicious use of pesticides; for example, using lower doses of aphicides on cabbages to allow natural predators to survive and control the remaining greenfly. Knowing the pest, the proper application of pesticides at the right time and in the right place can reduce the need to use pesticides and increase the operator's safety.

Another method of IPM includes developing the host plant's resistance to pests and pathogens (diseases). Probably three-quarters of all crops in industrialised countries have some form of resistance.

Autocidal controls are those which cause the pest to bring about its own downfall. The classic case is the screw-worm, an aptly named pest of livestock in the southern USA. Sterile males were mass released to mate with females. Since screw-worms only mate once before dying, the population decreased dramatically. Other successful examples of IPM can be found all over the world. Experimental farms in Cambridge, for instance, use kestrels to control rodents which attack sugar beet and vegetable seed beds. The Chinese recognise that pesticides can be hazardous to people, so they study the way insects breed, and the use of exclusion tactics and natural controls. Fish and ducks, for instance, are used to control pests in paddy fields.[7,8,9]

- *Join an organic gardening/farming association.* See lists on p. 234.
- *Keep your eyes open.* Whether you are interested in pests in your garden, or pesticides at work, you are in the best place to spot things going wrong and take preventive action early on.
- *Lice among children*: the incidence may be much higher than local, official figures suggest. Head lice appear to be increasing despite all the pesticides used to 'treat' them. A voluntary

body, the Community Hygiene Concern's (CHC), survey revealed that over one in four pupils catch head lice at least once a year. Other health areas have found only 2%. CHC considers it is the duty of authorities to regulate the use of medicinal insecticides, which are being used twice a month with some children.[10] Many of the insecticide treatments do not kill the eggs. The CHC is campaigning for an egg-killing treatment which does not contain any conventional pesticide. Every encouragement is given to parents who wash out the lice from their children's hair with vinegar and paraffin (see *vinegar*). Complete the cure by manually removing the eggs with a metal nit comb while the hair is still wet.

A co-ordinated, school-wide effort is the key to reduction. Following a Bug Busting Day, a school could aim for a 'no nit policy', but this should be very sensitively handled with *all* parents agreeing and being consulted and depending on age, perhaps to pupils too. The Isle of Man has become the UK's only 'lice-free zone', following a bug-busting campaign in 1986 when one in twenty children were found to be infected.

- *Mulching* keeps your weeds down. It involves covering the ground completely with some neutral or beneficial material. This stops weeds germinating or taking over by depriving them of light, space or nutrients. You can use straw or bark, or even plastic sheeting, old carpets and newspaper. When councils use bark as a mulch in public parks and gardens, it means they are not using herbicides on that plot.
- *No nicotine*. We know nicotine is organic, but it's still very nasty. Our advice is: don't smoke it or spray it.
- *Organic produce*: buying organic is the only way to know what pesticides are in your food. Some food sold as organic has been found to have pesticide residues. Fraud apart, even organic farmers cannot control pesticides from neighbours or in background pollution. There are several symbols to look for. The Soil Association, Demeter, Biodyn, the United Kingdom Register of Organic Foods (UKROFS), and Organic Farmers and Growers Ltd are all different yet broadly acceptable standards. Watch out for changes in standards. None are supposed to use pesticides. Organic farmers and growers ideally are inspected annually and have not used chemicals for at least two years. At some stage there will be an EC symbol –but probably with a lower standard.[11] At pre-

sent there are only 1–2,000 organic farms out of nearly a quarter of a million in the UK – which explains why 70% of organic produce sold here is imported.

- *Plants for pest control.* Several plants act as insect repellents. Rubbed lavender or lavender oil will repel mosquitoes and other pests, and strewn in cupboards repels carpet beetles, silverfish etc. Similarly, rosemary branches in wardrobes and drawers repel moths. Dried cloves or garlic will help keep weevils out of stored food without tainting it. Santolina repels moths, fleas and mosquitoes. Bay leaves can be put in cupboards to keep away silverfish, and put into cereal packets to repel insects. Roman soldiers used to bath in thyme water, ostensibly to give them courage but also to get rid of lice. Broom boiled in oil is another old remedy for lice.

Pyrethrum has become well known as a natural insecticide and has given rise to a whole series of synthetic imitations. The plant can be grown in pots which can be sunk into the ground next to vulnerable plants like cabbages; alternatively the flowers can be collected, dried, and made into powders or sprays. Other plants affect other pests. Marigold roots inhibit nematode worms. Horseradish leaves can be made into a fungicide. Mice and rats dislike the smell of peppermint.[12]

Plants have long been known to 'prefer' certain other plants. The reason may be that one plant exudes substances that deter predators from the other, or the scents may confuse pests, or one plant may produce a substance that the other needs. The success of 'companion planting', as it is called, depends on experience and a fair bit of folklore. But planting herbs among and around garden plots is generally beneficial. The Henry Doubleday Research Association gives good advice. Specific examples of companion planting are tomatoes with chives, onions, parsley and cabbages, but not with fennel and kohlrabi. In addition, cabbages like potatoes and celery but dislike strawberries.[13]

- *Read* Rachel Carson's *The Silent Spring* if you have any doubts. It was the first, but is still the most prophetic book on pesticides.
- *Scandinavian* countries have shown the way on pesticide-reduction policies. Sweden introduced a long-term plan in 1986 to minimise the environmental damage caused by pesticides: the average usage in 1981–5 was to be halved. An envi-

ronmental levy on pesticides was doubled in 1988. Only one-third of pesticides used in Sweden now are applied on farms, and they were banned from use in forests ten years ago. The range of products which can be used by non-professionals has been greatly reduced. By the end of the 1980s pesticide usage was down by 35%, and the country expects to achieve its target by 1995.

Some of Sweden's reduction can be put down to using stronger pesticides (more potent active ingredients), so we shouldn't be over-enthusiastic about Scandinavia as the way forward.[14] Nevertheless, Sweden's approach shows that improvements can be done. Where pesticides *are* used, the Swedes have invested in better professional use, and they have spent money and resources on reviewing all old pesticides.[15] Other countries are now following Sweden's lead. Denmark, too, has a similar policy. Now the Netherlands is following suit, and there is talk of a 50% reduction plan in Australia.[16] Why not the UK?

- *Traps* can be either simple or highly developed devices. Boards, rocks, sacks, plastic bags, or cardboard placed on the ground catch snails, slugs, and woodlice underneath them. Collect them up and destroy them in detergent and water. Grease bands have long been used to catch pests crawling up trees. Yellow boards which attract the whitefly are painted with glue and catch whitefly in glasshouses.

  Good old sticky flypaper may be unsightly, but you don't wonder where the dead flies have gone and what vapours are in the air. And you can hang them up in corners out of sight. Keep them above the children's 'grasp zone', though! You can also buy those electric traps which butchers shops use. They electrocute insects that come into contact with them and give them a quicker death.

  Other traps include physical barriers that act by abrasion or dehydrating effects. Traps are particularly good for slugs – although cleaning them out is a bit off-putting. Use upturned grapefruit or baked potato shells to catch them. Even better, sink a bottle into the soil and half fill it with either milk or watered-down beer. Slugs love it. So do wasps if you want to decoy them. Some jam and water in a jar will also catch them.

  A trap developed in Australian orchards keeps fruit flies and other insects at bay. It consists of an upturned 2-litre plastic bottle, sawn off at the neck. The top is welded back inside and bait – marmite for fruit flies, meat for blowflies – added. The flies fly up and over, but not out. Wash it out when it's full.[17]

- *Use your head* when searching for alternatives. There is no point in jumping out of the frying pan into the fire.
- *Vinegar* mixed with paraffin, washed into your hair and left for half an hour, can control head lice. It is better for you than regular treatments of Lindane, Malathion or Carbaryl. (See *Lice* for more information.)
- *Wood preservatives* should not really be used, as they treat the symptoms rather than provide the cure. 'Damp, decay, and beetle infestations are rarely inherent. Analysis of the symptoms is essential' say Hutton and Rostron, a group of architects who have used non-chemical methods in famous historic buildings. If you must use a chemical, use boron rods.

There are two beneficial fungi (*Trichoderma harzianum* and *Scytalidium* FY strain) that fight off decay fungi. In five-year tests they protected uncreosoted poles underground and extended the life of poles that had been creosoted. The US Forest Products Laboratory has developed a simple water repellent made from varnish, paraffin wax and mineral spirits. It has been used to protect decks, fences and furniture as successfully as the most toxic wood preservatives.[18]

- *Experiment* for yourself. Look to see what happens. Keep an eye on, or record, how much damage is caused by whatever pest is causing you problems. Note what works best. Talk to other people, especially 'old-timers'. Get hold of an old book on insect pests – published before synthetics were invented. You can find how life went on before. Look out for remedies which did not involve chemicals.

Mrs Eleanor Ormerod, who wrote at the turn of the century, is particularly fascinating.[19] In the summer of 1990, when researching and writing this book, one of your authors got rid of gooseberry sawfly by following her advice. She advises scraping off the top 2–3 inches of soil during the winter, which will get rid of all this pest's overwintering cocoons and thus destroy its life cycle. Then tell others of your success, just like we have.

- *You can help* bring down the price of food produced without pesticides. The price difference is huge and takes it way beyond what people on low or even medium incomes can afford. Help by writing to supermarkets to ask them to lower the price. Why can't they do 'loss leaders' (i.e. give a subsidy) to pesticide-free food? To subsidise the healthy option makes sense. Write to the Minister and your MP, too. Let them know you want more action on pesticides. Say how scandalised you were to learn that 6% of fruit, vegetables, cereals and cereal products were above the government's own Maximum Residue Levels in official tests done in 1988–89. Be politely firm. They wouldn't accept such failure from the car industry without much finger wagging at the employees!

- *Zootermopsis sp.* probably doesn't mean much to you. If we called it a damp wood termite you might appreciate that it is one of the worst insect pests in the world. It is found in Australia and the USA. Two potent pesticides, Chlordane and Heptachlor, are seen by many to be the only answer, and this is said to justify the continued clearance of these pesticides for other purposes. But there are alternatives. Much can be done at the time of building – and afterwards during maintenance – to reduce infestation. Removal of any wood debris, allowing sufficient slope for drainage, building concrete foundations, providing adequate ventilation and repairing faulty water systems will help.

Certain woods (such as the Douglas fir) offer resistance. There are nematode mixtures which search and enter termites, living off them and then going off to find another. One nematode, *Neoaplectana carpocapse*, got rid of 96% damp wood termites and 98% soil termites within three days in one Australian study. There are professional nematode operators in New York. Such treatment is bit more expensive, but it is safe, easy to use, leaves no toxic residues and does not lead to resistance or possible law suits later.[20,21,22]

Where to in the Future?

There is lots do. A sensible, general approach is to make pesticide use safer, if they must be used; to cut their use down; and to create and support alternatives far more than the current pathetic financial support. The previous chapter outlined some practical ideas, and now we offer you two visions of the future. The first is a technical one, in which biotechnology and genetic engineering are to be the answer to some of the problems this book has pointed out.

The technical fix vision is being invested in by many of the big companies. Powerful genetic modifications can produce biological pesticides. Companies are also breeding new strains of plants which are resistant to their own company's pesticide. Instead of breeding plants to be resistant to pests, more emphasis is being placed on breeding plants which are resistant to herbicides. The idea is that farmers will be able to use more of Brand X of pesticide if they plant Brand Y seed. Spray without a care is this dream. But will it mean more damage to the environment? Will we have to go through another generation of blunders?

And then there is another vision, which learns lessons from the past. This puts more emphasis on investing in alternatives to ham-fisted technologies. More emphasis is put on prevention and long-term ecological soundness. According to this human- and environment-centred vision, it is important to get the process of making decisions about chemicals right.

The technical fix vision

Genetic engineering tries deliberately to change the genes of an organism in order to alter one or more of its characteristics. Genetic engineering at present represents a small proportion of all the activities known as biotechnology.

There is great incentive for agrochemical companies to invest in the development of genetic engineering. Resistance of pest populations to chemical pesticides has been mentioned as a growing problem in a number of areas where pest control is carried out. The cost of developing a new pesticide is now put at £35–40 million, and it may take 8–12 years for the product to reach the market. The investment required to develop a genetically engineered pesticide has been estimated at £5 million, and the length of time to reach the market would be some three years.[1] The potential market by the end of the century,[2] for genetically engineered seeds has been estimated at £6.8 billion and for genetically engineered pesticides at £6–8 billion.[3]

In July 1989, the Royal Commission on Environmental Pollution released its thirteenth report, *The Release of Genetically Engineered Organisms to the Environment.*[4] The report is about the environmental issues raised by the release of such genetically engineered organisms (GEOs). It discusses

- the effects that release might have on the environment
- the procedures necessary to identify, assess and minimise any risks to the environment
- and the regulatory arrangements needed to ensure protection for the environment.

Risks to health are regarded as one of the dangers following unauthorised release. The Royal Commission recognised that the environmental impact of released organisms may be undesirable as well as beneficial. Although stating that: 'It will be prudent to begin with the assumption that an introduced gene could spread widely and then to challenge that assumption', the report does not consider that there should be a ban or moratorium on releases. Instead, 'We consider it essential that the release of GEOs is conducted from the outset under appropriate statutory control.'

A new Advisory Committee on Releases to the Environment (ACRE) has been set up under the authority of the Environment Protection Act 1990, to scrutinise releases. The law also includes the provision of a public register of applications for release, together with the recommendations of ACRE. This is an improvement on the controls over chemical pesticides.

Two topics of current interest in the application of genetic engineering techniques to pesticides are the development of herbicide resistance and insect resistance in plants.

Herbicide resistance

Herbicides kill weeds, but can also kill the crop they are supposed to protect. The crop plant itself can be genetically altered to be made tolerant, or resistant, to doses of the herbicide. This development could lead to an increased use of more toxic chemicals and greater risks to farm workers. Will there be

- greater environmental damage?
- more chemical residues in food?
- increased production costs?
- any danger of crop loss due to increased vulnerability of the variety?

In 1991 the Minister of Agriculture is set to give approval to a genetically modified sugar beet, resistant to glyphosate. The two companies behind the patent argue this is more economical and environment-friendly. Others like Genetics Forum, and ourselves are not so sure.[5]

Insect resistance

Insect-resistant plants are a further area where genetic engineering techniques have been employed in research. There is a view that, because of the availability of cheap pesticides, traditional plant-breeding techniques have been aimed at increasing yields rather than improving the resistant qualities of plant varieties to particular insect pests. These 'home-grown' resistant qualities come about because some plants contain within their structure substances with insecticidal effects which can be used in solution.

The bacteria *Bacillus thuringiensis (B.t.)* contains a toxin gene which makes *B.t.* an effective biological control agent. Research is already under way to engineer insecticidal plants using this gene. Field trials have involved the use of a genetically altered microbe which was inoculated into corn plants. According to one commentator, the *B.t.* toxin gene was then found in flea beatles in the field.[6] The process of gene transfer is not yet understood: safeguards are required to ensure that other plants, or insects, are not affected by this process. There is also concern that, if genetically engineered plants are marketed and grown on a large scale, then pest resistance and adaptation may also develop.

Releases that went wrong

Already there have been reports about controversial releases. In 1986, an American research organisation commissioned the Pan American Health Organisation to conduct an experiment on a herd of cattle in Argentina using a genetically engineered rabies vaccine. The virus used is said to have passed from the cows to their contacts, and in two cases to people involved in milking them.[7] The circumstances of the release and the question of whether it was or was not authorised are contentious. Nevertheless, the case serves to illustrate the problems that may occur in situations where guidelines have yet to be laid down, and in experiments that cross national boundaries.

In 1984, an American variety of insect-resistant potato had to be withdrawn from the market when it was found to be producing higher levels of the carcinogens solanin and chaconin than are normally found in ordinary potatoes.[8] Similarly, a new variety of insect-resistant celery was withdrawn when it caused dermatitis in production workers due to the high level of an irritant and carcinogen that it produced.[9]

Research

Most research will be carried out by corporations. In 1985, in the USA, 79% of biotechnology-related patents went to corporations, while the remainder went to universities and government agencies.[10] In the UK, government research councils have disagreed between themselves over the control of research in biotechnology partly because the field is developing so fast; the outcome could affect the freedom of researchers. The government is encouraging scientists to sell their research to industry, which sets a premium on commercial interest. 'The danger . . . is that basic research could play second fiddle to the demands of industry.'[11] The Research Councils, which are becoming ever shorter of funds, also seem to be encouraging scientists to collaborate with industry as research in biotechnology has been reported as 'a buyers' market'.[12] Therefore,

● will 'public good' research will be carried out in the private sector?
● will the development of biotechnology techniques continue to service current conventional methods of agriculture, thus requiring high chemical inputs?

Already large trans-national corporations are taking over small biotechnology companies. One estimate is that over half of the biotechnology companies in the agricultural sector expect to be taken over by 1995.[13] One important issue raised by genetic engineering is that these large companies want to patent life forms to recover their investment. The European Commission, at time of writing, is introducing a directive (law) to allow this.[14]

The human-centred vision

And now for what we think are the real priorities. How can *you* act on the information in this book? There are things you can do on your own, and things which can only be done with others. Some problems can be tackled locally, others only nationally or internationally. The European Consumers' Pesticide Charter (see p. 225) spells out the details of some of what we would like to see happen.

On your own

Standards can be tightened up. Keeping asking the questions. Keep writing letters. For instance, keep up pressure on your local supermarket manager to use the chain's buying power to lower pesticide use. Agrochemical companies will stop producing hazards if economic forces render them unprofitable. There are a lot of 'you', and relatively few of 'them'. But remember that sometimes manufacturers are not too bothered about dropping one pesticide, as they may have a new 'wonder product' that needs a market.[15] In saturated or stagnating markets, outside pressure can lead to innovation, to the benefit of companies.

Get any friends you have in other European countries to write a letter to the President of the European Commission, Berlaymont Building, 200 Rue de la Loi, Brussels B-1040, Belgium. Write to your MP with any of the questions we have not been able to answer in this book.

With others

The authors of this book co-produced and published the current major plan on how to cut pesticide dangers, the European Consumers' Pesticide Charter (see p. 225). Have a look at it.

Take it to any appropriate organisation or consumer group to which you belong, and ask for its support. The UK is part of Europe, so any progress should be Europe-wide. If improvements are not international, the whole idea of European harmonisation by the end of 1992 will be a nonsense. Equally, improvements usually come from people working together, not on their own.

Together, we can aim for better:

1. INVESTMENT

- *labelling of foods*
  Both pre- and post-harvest use of pesticides must be recorded. A new European P for Pesticide labelling scheme should be in place in 1995.
- *labelling of pesticide containers*
  This must be improved to include symptoms of poisoning, preventative measures, alternatives and where to complain.
- *training*
  Manufacturers should pay for training sessions for the public at garden centres where their products are sold. There should be regular retraining for all workers coming into contact with pesticides.
- *improved packaging*
  Manufacturers should also put money into research for improving containers and disposal techniques.
- *company levy*
  A fund to enable full and extensive monitoring of pesticide residues in food is needed. Such a levy should be part of the approved process.
- *better public investment*
  Decently funded research into alternatives is overdue. Spending should be vastly increased. MAFF claims that the Agricultural Development Advisory Service (ADAS) actively promotes Integrated Pest Management. Currently the government provides £1 million for research and development (R and D) in organic production. The total R and D programme for pests and pesticides, claimed to be aimed at reducing pesticide usage, amounted to £20 million in 1990. This compares with UK pesticide sales in excess of £1,000 million a year.

2. LAWS

- *new legal rights*

  The law should give the benefit of the doubt over safety in principle to the public and the environment. The presumption should be that any level of pesticide is dangerous, unless proved otherwise. The UK should consider adopting something like the US Delaney Clause, named after the US Congressman who chaired a committee which took evidence for eight years on food safety. The Delaney Clause stated that anything cancerous to animals should be assumed cancerous for humans too. We think the Delaney Clause, although withdrawn, needs modifying and updating for Europe. In spirit it was pro the public health.

- *more open decision-making*

  Key committees should be opened up. The Advisory Committee on Pesticides should include at least three independent consumer, environment and trade union representatives. There is now representation of this kind on some of its sub-committees, but not on the main one. Commercial confidentiality over data should be no excuse for secrecy. If the product is in public use, the data should be public, or no licence should be given. Wherever there is an assessment between risk and benefits, those who are at risk must be part of that assessment. Companies should 'put up or shut up'.

- *more and better enforcement*

  Better monitoring, improved resources for inspectors and prosecution for breaches of the law should be implemented to make any sense of pesticide standards.

- *campaign for the Fourth Hurdle*

  We think a pesticide should be judged, not just by how safe or effective or good quality it is, but also by whether it is really needed. Assessing that need is known as the Fourth Hurdle (see p. 86). Manufacturers don't want the Fourth Hurdle to be used in evaluating new technologies in Europe or globally. We think that unless new pesticides are judged by how needed they are, economic considerations will dominate. Why should only the manufacturers be allowed to judge how 'necessary' an invisible pesticide residue will be?

- *monitoring of exports*

  These should ensure that PIC is followed, and not avoided, now PIC is formal UN policy. Where chemicals have been removed for 'commercial reasons', they should be closely watched.

- *product liability laws*
  We would like to see much tougher product liability laws internationally. This book has argued that essentially you, the environment and users have been used as guinea pigs in an experiment. Tougher product liability laws would help workers, consumers and the public by giving them leverage on the companies which manufacture and market pesticides.

3. WORKING TOGETHER

- *consumer groups*
  Consumers could put pressure on retailers, processors and farmers to use fewer pesticides. They are already involved in such pressure, but need your help. The last resort, if there is no give, is to organise boycotts of some key products or of particularly badly behaved companies. There are already campaigns for pesticide-free chocolate, bananas and cotton. Consumers have the right not to be worried about pesticides. The Food Safety Act 1990 says food should be as you 'demand' it. Did you ask for pesticide residues?

- *environmental groups*
  Environmentalists could build up databases on pesticides. These would include not only data sheets, but details of incidents, prosecutions and contaminations. Already there are environmental databases stored electronically and available through electronic networks such as POPTEL.

- *trade unions*
  Worker organisations have conducted surveys of their members and are represented via the Trade Union Congress (TUC) on both sub-committees of ACP in MAFF and the various HSE committees. Council workers in London are making progress towards control of hazards from pesticides used in parks and paved areas. The action began in 1986 when unions in Islington, Haringay, Hammersmith and Fulham won the banning of chemicals whose dangers had been highlighted. Waltham Forest banned a pesticide product containing Amitrole, Atrazine, and Diuron. Islington banned 82 pesticide products. Policies gave unions a large say in the procedures for selecting, storing, transporting, applying and

disposing of pesticides.[16] Barnet's Recreation Committee agreed in July 1990 to ban weedkillers containing Atrazine and Simazine and substitute alternative products plus hand-weeding.[17]

● *a policy of reduced pesticide usage*
This would have to include education, incentive schemes, advice services, research and development of alternatives.

● *working together is best*
There need be no conflict among all these groups. Most trade union members are consumers. Most consumers live in an environment. Many environmental concerns are similar to those of trade unions. If there was any one issue that brought these groups together it is pesticides. We would like, more than anything, for all public interest groups to work together on pesticides. To pool scientists, advisers, campaigns and support makes sense.

You are part of the pesticide story

The story of pesticides never stops. It is continually changing. Look how the laws change all the time. The standards keep shifting. Pesticides come and go. There have been improvements in beating secrecy, in container design and in law. But as fast as there are improvements, other hazards enter the picture or pop up on the other side of the world, connected to this side.

When decisions are made on pesticides, someone is judging between benefits and risks. Pesticide safety is all about people, power and politics. Like it or not, the environment and your health are part of that power game. The experience of public interest groups over the last 20 years' work on pesticides is quite clear on this point.

Pesticide use is linked to food production. And food production changes all the time too. Changes in eastern Europe will bring changes in pesticide use. The major chemical producers are already aware of new opportunities in this region. The latest General Agreement on Tariffs and Trade (GATT) round alters the face of production and pesticide usage, as will the aftermath of the Gulf War.

Underlying these movements are some long-term issues. There is overproduction of food in both the USA and Europe. There is pressure from each to use less subsidy and to produce less food. Both USA and Europe 'dump' surplus food on the world, driving selling prices down in the rest of the world. Each family in Europe pays about £6 extra for its food each week to cover surpluses, storage and export subsidies, as this dumping is called.[18]

Food price wars put pressure on farmers, in both the developed and underdeveloped areas of the world, to produce as cheaply as possible. A vicious circle develops: more intensive use of the better land, abandoning poorer land and getting rid of agricultural labour. Replacing old skills with new chemicals is one element in this vicious circle.

What a wonderful world it would be if there was not this drive to overproduce. All land could be better used. More people could live and work in the country. Farmers would be happy. Farm workers need not be on dreadful and sometimes poverty wages. Pesticides would not be needed as much. There could be time to investigate pest problems locally. Options would open up because there wouldn't be all that pressure to increase and intensify production. The aim of any sane food policy would be, not to create mountains of food, or to use food as a weapon, but to feed people with good, wholesome, uncontaminated food, that they can be proud of and enjoy.

Funny old world. Help make it sane and safe.

Table 1

PART 2

What Pesticides Do You Find in Which Foods?

Pesticide residue analysis of foods has not been carried out on a large scale. MAFF has recently increased its analysis and has begun to publish results annually.

The following table lists pesticides used, pesticide residues analysed, and results during the production of foods in the 1980s. The majority of results have been published by MAFF. Many tests have been carried out by food retailers, but these results have not been made public.

To find out more about the pesticides listed, and their potential harmful effects, turn to the tables on pages 182 and 214. If you do not find the pesticides listed on either of these tables, it is because they have been banned (see Chapter 12), withdrawn or no longer cleared for use in the UK.

Each food group in the table is divided into four sections.

Pesticides used

This figure gives an approximate number of pesticides used in vegetable and crop production. The source for this estimate comes from either *The UK Pesticide Guide 1990* (published by CAB International and BCPC), or MAFF usage reports. These reports cover various aspects of the use of agricultural pesticides, and cover only England and Wales. The dates quoted vary, as does the most up-to-date information on the particular food group. Reports drawn on here include *Review of Usage of Pesticides 1980–1983* (Survey Report 41 (1986), ADAS), *Vegetables for Human Consumption 1986* (Survey Report 64 (1990), ADAS) and *Arable Farm Crops 1988* (Survey Report 78, Pesticides Usage (1990), ADAS). The number of pesticides used is not always given, as figures are not always available.

Extent of MAFF analysis

There are three main reports which give details of MAFF analysis: *Report of the Working Party on Pesticide Residues: (1982–85)* (Food Surveillance Paper No. 16, HMSO, 1986), *Report of the Working Party on Pesticide Residues: 1985–88* (Food Surveillance Paper No. 25, HMSO, 1989) and the MAFF and HSE *Report of the Working Party on Pesticide Residues: 1988–89* (Supplement to Issue No. 8, 1990, of the *Pesticides Register*, HMSO, 1990). Studies are listed in date order. The number of samples tested (if known) and the residues analysed (if known) are listed. Note that, for DDT: reference is made to 'total DDT', which indicates that analysis assessed both DDT and its breakdown products TDE and DDE. Lindane is analysed as gamma HCH. Lindane is sometimes known as Benzene hexachloride (BHC) or Hexachlorocyclohexane (HCH). Lindane is 99% or more of gamma HCH; there are other types of HCH (alpha, beta, delta – HCH) which are impurities and which are sometimes indicated.

Residues found

The results of these studies are also listed in date order. Where possible, the overall percentage of residues found, and the residues detected, are given. These results may indicate the residues which are likely to occur in food. But the sample sizes and the number of residues looked for are small. As an extreme example, during 1988–89 only one sample of UK retail lamb was analysed. The trend has been to increase the number of tests carried out, but it is making slow progress.

Maximum Residue Limits (MRLs)

Certain MRLs specified are under UK legislation and they are listed in this section. There are some references to European Community MRLs: these were in operation before the UK set MRLs by law. Remember that MRLs are not safety levels, but simply an indication of whether Good Agricultural Practice is being followed. The existence of a number of pesticide-commodity MRLs means that either there are many pesticides that can be used on that commodity, or that the use of pesticides on that crop is well patrolled by law. Refer to Chapter 10 for more about MRLs. Tables 2 and 3 indicate whether MRLs have been set for that pesticide.

Apples

Pesticides used

About 90 are regularly used on apples. The top five most extensively used on dessert, cooking and cider apples in 1983 were Triadimefon, Dithianon, Captan, Fenarimol and Binapacryl (since banned).

Extent of MAFF analysis

1. Between 1981 and 1984 an apple sample of unknown size was analysed for organochlorines (unspecified), organophosphates (unspecified), Binapacryl, Biphenyl, Captan, Carbendazim, Diphenylamine, Dithianon, Ethoxyquin, 2-Phenylphenol, pyrethroids, Thiabendazole and Vinclozolin.
2. Between 1983 and 1984 62 samples of apples were tested for DDT and its breakdown products.
3. Between 1985 and 1986 46 samples of UK apples were tested for Carbendazim and Metalyxyl, and 96 samples of imported apple were analysed for Diphenylamine, Captan, Carbendazim and Metalyxyl.
4. Between 1988 and 1989 apple samples were analysed for 37 pesticide residues, including Azinphos-ethyl, Azinphos-methyl, Bitertanol, Buprimate, Captan, Carbaryl, Carbendazim, Chlorpyrifos, Cyfluthrin, Cypermethrin, DDT, Deltamethrin, Diazinon, Dimethoate, Dinocap, Diphenylamine, Dithianon, dithiocarbamates, Endosulfan, Ethion, Ethoxyquin, Fenarimol, Fenitrothion, Fenvalerate, Flucythrinate, gamma HCH, Imazalil, Metalaxyl, Parathion, Parathion-methyl, Permethrin, Phenthoate, Tetradifon, Thiabendazole and Triadimefon.

Residues found

1. The 1981–4 study: one sample exceeded the MRL Azinphos-ethyl. Residues found below the MRL are not published.
2. The 1983–4 study: 6% of samples contained DDT.
3. The 1985–6 study: 50% of samples contained residues (80% UK and 35% imported). In the UK samples Carbendazim was found, and in the imported samples Diphenylamine and Ethoxyquin were detected.
4. The 1988–9 study: 52% of samples contained residues

(51% UK and 53% imported). Detected were Azinphos-methyl, Carbendazim, Chlorpyrifos, Dimethoate, Dipheny-lamine, dithiocarbamates, Metalaxyl, Phosalone, Thiaben-dazole and Vinclozolin.

Maximum Residue Limits

Apples have MRLs for 36 pesticides.

Barley

Pesticides used

At least 85 are used on barley. On spring barley in 1988 the most extensively used insecticides were Lindane, Dimethoate and Demeton-s-methyl; fungicides Tridemorph and Propi-conazole; and herbicides Ioxynil, Mecoprop and Bromoxynil.

Extent of MAFF Analysis

1. In 1982 one sample of retail barley was analysed for Lin-dane (gamma-HCH) and its impurities (alpha and beta-HCH), and DDT and its breakdown product DDE (most barley is used for animal feed and malting, so little is sold retail).
2. In 1984 114 'feeding stuff' samples were analysed for unspecified organophosphates and organochlorines.

Residues found

1. The 1982 study: alpha-HCH was found in one sample.
2. The 1984 study: residues of (5%) alpha-HCH, (17%) beta-HCH, (43%) gamma-HCH, (62%) Pirimiphos-methyl, (4%) Chlorpyrifos-methyl, (6%) Malathion and (1%) Fenitrothion were found.

Maximum Residue Limits

Barley has MRLs for 31 pesticides.

Beans

Pesticides used

At least 39 different pesticide types are used on peas and beans. In 1988 the most extensively used insecticides were Cypermethrin, Pirimicarb, Dimethoate and Triazophos; fungicides Chlorthalonil and Carbendazim, and Thiabendazole/Thiram for fungicide seed treatment; and herbicides Terbuthylazine/Terbutryn, Simazine, Cyanazine and Simazine/Trietazine.

Extent of MAFF analysis

1. Between 1981 and 1984 a bean sample of unknown size was analysed for organochlorines (unspecified), organophosphates (unspecified), Captan, Carbendazim and dithiocarbamates (five active ingredients).
2. Between 1988 and 1989 26 broad bean samples were analysed for four pesticide residues: Carbendazim, Malathion, Dimethoate and Pirimicarb; 20 runner bean samples were analysed for the same four residues plus demeton-s-methyl.

Residues found

1. The 1981–4 study: residues found below the MRLs have not been published.
2. The 1988–9 study: no residues were found.

Maximum Residue Limits

Beans have MRLs for 29 pesticides.

Beef

Pesticides used

The most likely sources of pesticide contamination are feed and water. The most frequent culprits are fat-soluble, for example Lindane and Dieldrin.

Extent of MAFF analysis

1. Between 1981 and 1983 27 samples of imported meat products were analysed for Lindane (gamma-HCH) plus its impurities alpha and beta-HCH and DDT plus its breakdown products DDE and TDE. The beef products analysed were corned beef, beef extract and beef fat.
2. Between 1984 and 1986 140 samples of UK and imported retail beef were analysed as part of a survey of meat and poultry products. The pesticides analysed were organochlorines (eight specified pesticides) and organophosphates (up to 22 unspecified pesticides).

3. Between 1985 and 1987 139 samples of meat products were analysed for organochlorines (10 specified pesticides) and organophosphates (up to 22 unspecified pesticides).
4. Between 1988 and 1989 27 samples of beef (7 UK, 14 imported and 6 of unknown origin) were analysed for 23 pesticides including Aldrin, Bromophos, Carbophenothion, Chlorfenvinphos, Chlorpyrifos, cis-Chlordane, trans-Chlordane, DDT and its breakdown products, Diazinon, Dieldrin, Endrin, Lindane and its impurities Heptachlor, Heptachlor epoxide, Hexachlorobenzene, Oxychlordane and Propetamphos.

Residues found

1. The 1981–3 study: 6% of corned beef and 80% of beef fat contained residues of DDT breakdown products.
2. The 1984–6 study: 13% of UK and 30% of imported samples contained residues of HCB, Lindane and DDT breakdown products and Dieldrin.
3. The 1985–7 study: 1% of corned beef contained residues of DDE, and 14% of tongue contained residues including Lindane and DDT breakdown products.
4. The 1988–9 study: 15% of samples contained residues of Dieldrin and Lindane.

Maximum Residue Limits

MRLs for beef come under 'products of animal origin' and in particular 'meat, fat and preparations of meat'. There are 14 MRLs for this category.

Brussels sprouts

Pesticides used

In 1986, the most widely used insecticides were Demeton-s-methyl, Cypermethrin, Deltamethrin, Chlorfenvinphos and Pirimicarb; fungicides Iprodione, Benomyl, Triadimefon and Chlorothalonil; herbicides Propachlor and Trifluralin. This refers also to broccoli, kale, cabbage, cauliflowers, calabrese, radishes, turnips and swedes.

Extent of MAFF analysis

1. Between 1981 and 1984 a sample of brussels sprouts of

unknown size was analysed for organochlorines and organophosphates (both groups unspecified).

Residues found

1. The 1981–4 study: residues found below the MRLs have not been published.

Maximum Residue Limits

Brussels sprouts have MRLs for 30 pesticides.

Cabbages

Pesticides used

See *Brussels sprouts*. For Chinese cabbage, see *Lettuce*. Standard cabbage types are Savoy, autumn/summer, spring, white and winter.

Extent of MAFF analysis

1. Between 1981 and 1984 a sample of cabbages of unknown size was analysed for organochlorines, Benomyl, Thiabendazole (in green cabbages) and Carbendazim and Iprodione (in white cabbages).
2. Between 1983 and 1984 122 samples were tested for DDT and its breakdown products.
3. Between 1985 and 1986 15 samples of white cabbage were tested for post-harvest treatment residues Carbendazim and Iprodione.
4. Between 1985 and 1987 323 samples of cabbage were analysed for DDT and its breakdown products.
5. Between 1988 and 1989 47 samples of green cabbage were sampled for 19 pesticide residues including Aldicarb, gamma-HCH, Pirimiphos-methyl, Tetradifon, Triazophos, Diflubenzuron, Demeton-s-methyl, Disulfoton, Phorate, Cypermethrin, Deltamethrin, Dimethoate, Carbofuran, Azinphos-methyl, Chlorpyrifos, DDT and its breakdown products, Thiometon, Trichlorfon and Quinalphos. 25 samples of Chinese cabbage (i.e. the salad vegetable, not cabbages imported from China) were also analysed for 21 pesticide residues including Carbendazim, Carbaryl, Chlorothalonil, Cypermethrin, DDT and its breakdown products, Deltamethrin, Demeton-s-methyl, Dimethoate,

dithiocarbamates, Endosulfan, Hexachlorobenzene, Lindane, inorganic bromide, Iprodione, Malathion, Metaloxyl, Permethrin, Pirimicarb, Quintozene, Tolclofos-methyl and Vinclozolin.

Residues found

1. The 1981–4 study: results found below the MRLs have not been published.
2. The 1983–4 study: 16% of samples contained DDT.
3. The 1985–6 study: 33% contained Carbendazim and 40% contained Iprodione.
4. The 1985–7 study: 3% contained detectable DDT residues (of those studied in 1985, 4% contained DDT levels above the EC MRL).
5. The 1988–9 study: no residues were found in the green cabbage samples. 80% of Chinese cabbage samples contained residues (mostly inorganic bromide but also Pirimicarb).

Maximum Residue Limits

Cabbage has MRLs for 32 pesticides.

Carrots

Pesticides used

In 1986 the most widely used insecticide was Triazophos; fungicides Iprodione, Thiram, Mancozeb and Metalaxyl; herbicides Linuron and Metoxuron.

Extent of MAFF analysis

1. Between 1981 and 1984 a carrot sample of unknown size was analysed for organophosphates (unspecified), Benomyl and Thiabendazole.
2. Studies carried out by the National Vegetable Research Station between 1983 and 1984 found Carbofuran in four samples, Disulfoton in 17 samples and Phorate in 20 samples (Suett, D.L., (1986) 'Insecticide Residues in Commercially Grown Quick-maturing Carrots', Food Additives and Contaminants 3 (4):371-175, 1986.)

Residues found

1. The 1981–4 study: residues found below the MRLs have not been published.

2. The 1983–4 study: 32% samples contained residues of Phorate and Disulfoton (Suett 1986).

Maximum Residue Limits

Carrots have MRLs for 30 pesticides. Of these only one, Iprodione, is included in the top five used on this vegetable.

Cauliflowers

Pesticides used

See *Brussels sprouts.*

Extent of MAFF analysis

1. Between 1981 and 1984 a cauliflower sample of unknown size was analysed for organophosphates and organochlorines (both groups unspecified).
2. Between 1988 and 1989 38 samples of UK produce and 13 samples of imported produce were sampled for 19 pesticide residues including Aldicarb, gamma-HCH, Pirimiphosmethyl, Tetradifon, Triazophos, Diflubenzuron, Demeton-s-methyl, Disulfoton, Phorate, Cypermethrin, Deltamethrin, Dimethoate, Carbofuran, Azinphos-methyl, Chlorpyrifos, total DDT, Thiometon, Trichlorfon and Quinalphos.

Residues found

1. The 1981–4 study: residues found below the MRLs have not been published.
2. The 1988–9 study: no residues were found.

Maximum Residue Limits

Cauliflowers have MRLs for 30 pesticides.

Celery

Pesticides used

See *Carrots.*

Extent of MAFF analysis

1. Between 1981 and 1984 a sample of celery of unknown size was analysed for organochlorines (unspecified), organophos-

phates (unspecified), Chlorothalonil, dithiocarbamates (five pesticides), Iprodione, Pyrethroids and Vinclozolin.

Residues found

1. The 1981–4 study: residues found below the MRLs have not been published.

Maximum Residue Limits

Celery has MRLs for 28 pesticides, of which only one, Iprodione, is included in the *Pesticides used* section.

Chicken

Pesticides used

The most likely sources of pesticide contamination are feed and water. The most frequently culprits are fat-soluble, for example Lindane and Dieldrin.

Extent of MAFF analysis

1. Between 1984 and 1986 130 chicken samples were analysed for organochlorines (unspecified), organophosphates (unspecified; analysis not clear) and others (unspecified; analysis not clear).

Residues found

1. The 1984–5 study: 30% of UK and 13% of imported samples contained residues of HCB, gamma-HCH, and Dieldrin.

Maximum Residue Limits

MRLs for chicken come under 'products of animal origin' and in particular 'meat, fat and preparations of meat'. There are 14 MRLs for this category.

Citrus fruit

Pesticides used

Citrus fruit is not grown to any great extent in the UK. But, pesticides used during production include: Chlorpyrifos, Fenitrothion, Parathion-methyl, Ethion and 2 Phenylphenol.

Extent of MAFF analysis

1. Between 1981 and 1984 a sample of citrus fruit of unknown size (grapefruits, lemons, limes, mandarins, oranges, pomelos and uglis) was analysed for organophosphates (unspecified), sec-Butylamine, Biphenyl, Carbendazim, 2,4-D, Imazalil, 2-Phenylphenol and Thiabendazole. A sample of oranges of unknown size was analysed for an unknown number of pesticides residues.

2. Between 1988 and 1989 25 grapefruits and 13 lemons were analysed for 22 pesticide residues including 2-Aminobutane, Biphenyl, Carbendazim, Chlorpyrifos, Chlorpyrifos-methyl, Cypermethrin, 2,4-D, Dicofol, Dimethoate, Ethion, Fenitrothion, Fenvalerate, Imazalil, Malathion, Mecarbem, Methidathion, Omethoate, Parathion-methyl, 2-Phenylphenol, Pirimiphos-methyl, Tetradifon and Thiadendazole.

Residues found

1. The 1981–4 study: two grapefruit samples exceeded the MRLs for Chlorfenvinphos and Malathion; five lemon samples exceeded the MRLs for Azinphos-methyl, Chlorofenvinphos (x2), 2-Phenylphenol and Parathion-methyl; one mandarin sample exceeded the MRL for Chlorfenvinfos; three orange samples exceeded the MRLs for Chlorfenvinphos, Fenithrothion and 2-Phenylphenol. Residues below the MRLs have not been published.

2. The 1988–9 study: 84% of grapefruits and 38% of lemons contained pesticide residues. The grapefruits contained Biphenyl, Chlorpyrifos, 2,4-D, Ethion, Methidathion, 2-Phenylphenol and Thiabendazole. The lemons contained Biphenyl, Fenitrothion, Parathion-methyl, 2-Phenylphenol and Thiabendazole.

Maximum Residue Limits

Citrus fruit have 35 MRLs.

Cucumbers

Pesticides used

The five pesticides most extensively used, by treated area, between 1981 and 1982 were Imazalil, Iprodione, Benomyl, Pirimicarb and Cyhexatin.

Extent of MAFF analysis

1. Between 1981 and 1984 a sample of cucumbers of unknown size was analysed for residues of organochlorines (unspecified), organophosphates (unspecified), Chlorothalonil, Iprodione, pyrethroids and Vinclozolin.

Residues found

1. The 1981–4 study: residues found below the MRLs have not been published.
2. From Canadian analysis, residues of Dicofol and Endosulfan have been found in greenhouse and outdoor cucumbers (Food Additives and Contaminants 1990, Vol. 7, No. 4, pp. 545–54).

Maximum Residue Limits

Cucumbers have MRLs for 32 pesticides.

Grapefruit
See *Citrus fruit.*

Lamb

Pesticides used

The most likely sources of pesticide contamination are feed, water and sheep dips. The most frequent are fat-soluble pesticides, for example Lindane and Dieldrin.

Extent of MAFF analysis

1. During 1984 samples of fleece collected from 2,853 sheep were analysed for Dieldrin.
2. Between 1984 and 1986 335 samples of retail lamb were analysed for organochlorines (eight specified pesticides) and organophosphates (up to 22 unspecified pesticides).
3. Between 1984 and 1986 (the continuation of study 1), 11,373 sheep fleeces were analysed for organochlorines and organophosphates (both groups unspecified).
4. Between 1988 and 1989 one sample of UK lamb and 69 samples of imported lamb were analysed for 24 pesticide residues including Aldrin, Bromophos, Carbophenothion,

Chlorfenvinphos, Chlorpyrifos, cis-Chlordane, trans-Chlordane, DDT and its breakdown products, Diazinon, Dieldrin, Endrin, Lindane and its impurities, Heptachlor, Heptachlor epoxide, Hexachlorobenzene and Oxychlordane.

Residues found

1. The 1984 study: Dieldrin residues were found in 0.5% of samples.
2. The 1984–6 study: 40% contained residues (36% UK and 48% imported). Residues found were HCB (UK only), gamma-HCH (UK and imported), DDE (UK and imported), DDT (UK only), and Diazinon (UK only).
3. The 1984–7 study: during sheep fleece testing the following residues were detected: gamma-HCH, Dieldrin, Diazinon and Propetamphos.
4. The 1988–9 study: 28% of imported samples contained residues of Diazinon, DDT breakdown products and Lindane impurities.

Maximum Residue Limits

MRLs for lamb come under 'Products of animal origin' and in particular 'meat, fat and preparations of meat'. There are 14 MRLs for this category.

Leeks

Pesticides used

In 1986 the most widely used insecticide was Aldicarb; fungicides Iprodione, Chlorothalonil, Mancozeb/Metalaxyl, Thiram and Benomyl; herbicides Ixoynil, Chlorbufam/Chloridazon and Prophachlor.

This refers also to onions.

Extent of MAFF analysis

No information available.

Residues found

No information available.

Maximum Residue Limits
Leeks have MRLs for 28 pesticides.

Lemons
See *Citrus fruit*.

Lettuce

Pesticides used
About 30 pesticides are used on lettuce. In 1986, the most widely used insecticide was Demeton-s-methyl; fungicides Thiram, Iprodione and Vinclozolin; herbicide Metamitron. This refers also to usage on artichokes, asparagus, beetroot, Chinese cabbage, celeriac, chicory, courgettes, herbs, horseradish, marrows, pumpkins, rhubarb, spinach and sweetcorn.

Extent of MAFF analysis
1. Between 1981 and 1984 a sample of lettuce of unknown size was analysed for organochlorines (unspecified), organophosphates (unspecified), Chlorothalonil, dithiocarbamates (five pesticides), Iprodione, pyrethroids and Vinclozolin.
2. Between 1983 and 1984 58 samples of lettuce were analysed for DDT and its breakdown products.
3. Between 1988 and 1989 101 lettuce samples (77 UK and 24 imported) were analysed for 21 pesticide residues including Carbaryl, Carbendazim, Chlorothalonil, Cypermethrin, DDT and its breakdown products, Deltamethrin, Demeton-s-methyl, Dimethoate, dithiocarbamates, Endosulfan, Hexachlorobenzene, gamma-HCH, inorganic bromide, iprodione, malathion, Metalaxyl, Permethrin, Pirimicarb, Quintozene, Tolclofos-methyl and Vinclozolin.

Residues found
1. The 1981–4 study: residues found below the MRLs have not been published.
2. The 1983–4 study: 5% of samples contained residues of DDT.
3. The 1988–9 study: 100% of samples contained inorganic bromide residues; 63% of samples contained other

detectable residues: Carbendazim, Cypermethrin, Demeton-s-methyl, Dimethoate, dithiocarbamates (five pesticides), Iprodione, Metalaxyl, Pirimicarb, Tolclofos-methyl and Vinclozolin.

Maximum Residue Limits

Lettuce has MRLs for 30 pesticides.

Milk

Pesticides used

The most likely sources of pesticide contamination are feed and water. The most frequent are fat-soluble, for example Lindane and Dieldrin.

Extent of MAFF analysis

1 and 2. During 1981 and between 1984 and 1985 milk was tested as part two separate Total Diet Surveys. The following residues were analysed: organochlorines (unspecified), organophosphates (unspecified; analysis not clear) and others (unspecified; analysis not clear).

3. During 1988 120 samples of UK milk were analysed for 30 pesticides residues including Aldrin, Bromophos, Bromophos-ethyl, Carbophenothion, Chlordane, Chlorfenvinphos, Chlorpyrifos, Chlorpyrifos-methyl, Crotpxyphos, DDT and breakdown products, Diazinon, Dichlorvos, Dieldrin, Endrin, Ethion, Famphur, Fenitrothion, Fenthion, Lindane and impurities, Heptachlor, Hexachlorobenzene, Iodophenphos, Malathion, Phosmet, Pirimiphos-methyl, Propetamphos and Tetrachlorvinphos.

4. During 1989 126 samples of milk were analysed for 45 pesticide residues. They included those sought during 1988 (except Tetrachlorvinphos) plus Ametryn, Atrazine, Cyanazine, Cypermethrin, Deltamethrin, Fenvalerate, delta-HCH, Heptachlor epoxide, Permethrin, Prometon, Prometryn, Propazine, Simazine, Simetryn, Terbuthyalazine, Trietazine and Tetramethrin.

Residues found

1. The 1981 study: Lindane and its impurities, DDT breakdown products and Dieldrin were found.

2. The 1984–5 study: DDT breakdown products were found.
3. The 1988 study: 44% of samples contained residues including Hexachlorobenzene, Lindane and its impurities and DDT breakdown products.
4. The 1989 study: 12% of samples contained residues including Dieldrin, Lindane and its impurities, and DDT breakdown products.

Maximum Residue Limits

Milk has 15 MRLs.

Mushrooms

Pesticides used

In 1982 the following were extensively used: insecticides pyrethroids, (unspecified) Pirimiphos-methyl, Lindane and impurities, Dichlorvos and Diazinon; fungicides Thiabendazole, sodium chlorate and Benomyl; soil sterilant formaldehyde.

Extent of MAFF analysis

1. Between 1981 and 1984 a sample size of mushrooms of unknown size was analysed for organochlorines (unspecified), organophosphates (unspecified), Carbendazim, Chlorothalonil, dithiocarbamates, Iprodione, Thiabendazole and Vinclozolin.
2. Between 1983 and 1984 125 samples of mushrooms were analysed for DDT and its breakdown products.

Residues found

1. The 1981–4 study: the EC DDT and Dichlorvos MRLs were exceeded in two samples respectively. Residues found below the MRLs have not been published.
2. The 1983–4 study: 2% of the samples contained DDT.

Maximum Residue Limits

Mushrooms have 26 MRLs.

Oilseed Rape

Pesticides used

At least 65 pesticides are used. The last decade has seen a dramatic 550% increase in the area of oilseed rape grown, from

55,110 hectares in 1977 to 304,144 in 1988. It is mostly used for animal feed. During 1988 the major insecticides used were pyrethroids; fungicides Prochloraz, Carbendazim and Iprodione; herbicides Fluazifop-p-butyl, Propyzamide and Clopyralid.

Extent of MAFF analysis

1. Between 1988 and 1989, 80 animal feed rape samples were analysed for 15 pesticide residues including Aldrin, cis-Chlordane, trans-Chlordane, DDT and its breakdown products, Endosulfan, Endrin, Heptachlor, Heptachlor epoxide, Methoxychlor and Oxychlordane.

Residues found

1. The 1988–9 study: 5% of samples contained Lindane (gamma-HCH) residues.

Maximum Residue Limits

Oilseed rape has no MRLs.

Onions

Pesticides used

See *Leeks.*

Extent of MAFF analysis

1. Between 1981 and 1984 a sample of onions of unknown size was analysed for organochlorines and organophosphates (both groups unspecified).
2. Between 1988 and 1989 46 samples of bulb onions were analysed for eight pesticide residues including Aldicarb, Carbendazim, Carbofuran, Chlorpyrifos, Iodofenphos, Maleic hydrazide, Oxamyl and Triazophos.

Residues found

1. The 1981–4 study: residues found below the MRLs have not been published.
2. The 1988–9 study: 28% contained residues of Maleic hydrazide (42% UK and 14% imported).

Maximum Residue Limits

Onions have 28 MRLs.

Oranges
See *Citrus fruit*.

Pears

Pesticides used

In 1983 the five pesticides most extensively used were Dithianon, Captan, Amitrole/ammonium sulphamate, Simazine and Amitraz.

Extent of MAFF analysis

1. Between 1981 and 1984 a sample of pears of unknown size was analysed for organochlorines and organophosphates (both groups unspecified).
2. Between 1985 and 1986 48 samples were analysed for Carbendazim, Captan, Diphenylamine, Ethoxyquin, Metalazyl, 2-Phenylphenol, Thiabendazole and Vinclozolin.

Residues found

1. The 1981–4 study: residues found below the MRLs have not been published.
2. The 1985–6 study: of UK pears, 18% contained Carbendazim and 36% Vinclozolin; of imported pears, 4% contained Diphenylamine and 12% Ethoxyquin.

Maximum Residue Limits

Pears have 36 MRLs.

Peas

Pesticides used
See *Beans*.

Extent of MAFF analysis

1. Between 1988 and 1989 34 samples of in-pod peas were analysed for 11 pesticide residues including Azinphosmethyl, Carbendazim, Demeton-s-methyl, Dimethoate, dithiocarbamates, Fenitrothion, Malathion, Metalaxyl, Parathion, Pirimicarb and Triazophos.

Residues found

1. The 1988–9 study: no residues were found.

Maximum Residue Limits

Peas have 29 MRLs, of which only Dimethoate and Captan are included in the *Pesticides used* section.

Pork

Pesticides used

The most likely sources of pesticide contamination are feed and water. The most frequent are fat-soluble for example, Lindane and Dieldrin.

Extent of MAFF analysis

1. Between 1981 and 1983 25 pork samples were analysed for alpha-HCH, beta-HCH, gamma-HCH and DDT (plus its breakdown products DDE and TDE).
2. Between 1984 and 1986 161 samples of UK retail pork were analysed for organochlorines (eight specified pesticides) and organophosphates (up to 22 unspecified pesticides).
3. Between 1985 and 1987 80 samples of ham and 78 of processed pork were analysed for organochlorines (10 specified pesticides) and organophosphates (up to 22 unspecified pesticides).
4. Between 1988 and 1989 (A) 25 samples of ham, 35 of luncheon meat and 60 of bacon were analysed for 10 pesticide residues including Chlordane, DDT and its breakdown products, Dieldrin, Endrin, Lindane (gamma-HCH) and its impurities (alpha and beta-HCH), Heptachlor, Heptachlor epoxide and Hexachlorobenzene.
5. Between 1988 and 1989 (B) 36 samples (including six UK) were analysed for 23 pesticides including Aldrin, Bromophos, Carbophenothion, Chlorfenvinphos, Chlorpyrifos, cis-Chlordane, trans-Chlordane, DDT and its breakdown products, Diazinon, Dieldrin, Endrin, Lindane (gamma-HCH) and its impurities (alpha and beta-HCH), Heptachlor, Heptachlor epoxide, Hexachlorobenzene, Oxychlordane and Proetamphos.

Residues found

1. The 1981–3 study: Samples contained residues including Lindane and its impurities and DDT and its breakdown products.
2. The 1984-6 study: 13% of UK samples contained residues including HCB, Lindane and DDT and its breakdown products.
3. The 1985–7 study: 3% of ham samples and 23% of processed pork samples contained residues including Lindane and its impurities and DDT and its breakdown prodcuts.
4. The 1988–9 study (A): 8% of samples contained residues including Lindane and DDT breakdown products and Hexachlorobenzene.
5. The 1988–9 study (B): 20% of UK and 3% imported samples contained residues of pp'DDE.

Maximum Residue Limits

MRLs for pork come under 'products of animal origin' and in particular 'meat, fat and preparations of meat'. There are 14 MRLs for this category.

Potatoes

Pesticides used

At least 38 pesticides are used on potatoes. In 1988, the most extensively used insecticides were Pirimicarb and Demeton-s-methyl; fungicides dithiocarbamates (unspecified) and Fentin (unspecified) compounds; fungicides for seed treatment Thiabendazole and Tolclofos-methyl; herbicides Paraquat, Monolinuron and Linuron; desiccants sulphuric acid and Diquat.

Extent of MAFF analysis

1. Between 1981 and 1984 a sample of potatoes of unknown size was analysed for sec-Butylamine, Captafol, Chlorpropham, Dichlorophen, Dinoseb (since withdrawn), Tecnazene and Thiabendazole.
2. Between 1985 and 1986 67 samples of potatoes were analysed for 2-Aminobutane, Chlorophroham, Tecnazene and Thiabendazole.
3. During 1987 120 potato samples were analysed for: organochlorines (unspecified), organophosphates (unspecified

up to 70), Captafol, 2-Aminobutane, Carbendazim, Thiabendazole, Dichlorophen, Chlorpropham and Dinoseb.

4. During 1988 120 samples of UK (83), imported (22) and unknown origin (15) potatoes were analysed for 19 pesticide residues including 2-Aminobutane, Captafol, Carbaryl, Carbendazim, Chlorpropham, Chlorpyrifos, Diazinon, Dichlorophen, Dieldrin, Dinoseb, Fenitrothion, Lindane and its impurities. Hexachlorobenzene, inorganic bromide, Pirimiphos-methyl, DDT and its breakdown products, Tecnazene, Thiabendazole and Triazophos.

5. During 1989 126 samples of UK (93), imported (19) and unknown origin (14) potatoes were analysed for 20 pesticide residues. They included those sought during 1988 plus 2,3,5,6-Tetrachloroaniline and 2,3,5,6-Tetrachlorothioanisole.

Residues found

1. The 1981–4 study: one sample contained Captafol above the international MRL. Residues found below the MRLs have not been published.

2. The 1985–6 study: 4% contained 2-Aminobutane, 13% contained Chlorpropham, 76% contained Tecnazene and 19% contained Thiabendazole.

3. The 1987 study: 63% of maincrop potatoes and 40% of new potatoes contained Dieldrin, DDE, 2-Aminobutane, Chlorpropham, Tecnazene, Thiabendazole, total bromine in maincrop potatoes; and Dieldrin, DDT and its breakdown products, Tecnazene and total bromine in new potatoes.

4. The 1988 study: 44% contained residues (41% UK and 50% imported) including 2-Aminobutane, Dieldrin, Tecnazene, Thiabendazole, inorganic bromide and Lindane.

5. The 1989 study: 42% contained residues (43% UK and 50% imported), including Chlorpropham, Chlorpyrifos, DDT breakdown products, Dieldrin, Lindane and its impurities, inorganic bromide, Tecnazene, 2,3,5,6-Tetrachloroaniline, 2,3,5,6-Tetrachlorothioanisole and Thiabendazole.

Maximum Residue Limits

Potatoes have 35 MRLs.

Strawberries

Pesticides used

Some 70 pesticides are used on strawberries including: Benomyl, Carbendazim, Demeton-s-methyl, chlorpropham and Alloxydim-sodium.

Extent of MAFF analysis

1. Between 1981 and 1984 a sample of strawberries of unknown size was analysed for organochlorines (unspecified), organophosphates (unspecified), Benomyl, Chlorothalonil, Dichlofluanid, dithiocarbamates and Iprodione.
2. Between 1983 and 1984 72 samples were tested for DDT and its breakdown products.
3. Between 1988 and 1989 2 samples were analysed for 23 pesticides including Azinphos-methyl, Bupirimate, Carbendazim, Chlorothalonil, Chlorpyrifos, Cypermethrin, DDT and its breakdown products, Deltamethrin, Demeton-s-methyl, Dichlofluanid, Dicofol, Dimethoate, dithiocarbamates, Endosulfan, gamma-HCH, Iprodione, Malathion, Permethrin, Pirimicarb, Quinomethionate, Tetradifon, Triazophos and Vinclozolin.

Residues found

1. The 1981–4 study: two samples contained Dithiocarbamate residues above the EC MRL.
2. The 1983–4 study: 1% of samples contained DDT.
3. The 1988–9 study: 38% of samples contained residues (40% UK and 36% imported).

Maximum Residue Limits

Strawberries have 30 MRLs.

Swedes

Pesticides used

See *Brussels sprouts*.

Extent of MAFF analysis

There is no published data.

Residues found
There is no published data.

Maximum Residue Limits
Swedes have 29 MRLs.

Sweetcorn (Maize)

Pesticides used.
See *Lettuce.*

Extent of MAFF analysis
1. During 1981 about 330 samples of imported sweetcorn were analysed for over 30 unspecified organochlorines and organophosphates. About 160 samples were tested for an unspecified number of fumigants.

Residues found
1. The 1981 study: the overall percentages were not published, but OPS and OCS found were Dieldrin, Lindane, Malathion and Pirimiphos-methyl; fumigants carbon tetrachloride and carbon disulphide.

Maximum Residue Limits
Sweetcorn has no MRLs.

Tomatoes

Pesticides used
The five most used pesticides in terms of area in 1980–3 were Iprodione, Dichlofluanid, Vinclozolin, formaldehyde and Benomyl.

Extent of MAFF analysis
Between 1981 and 1984 a sample of tomatoes of unknown size was analysed for (2) organochlorines (unspecified), organophosphates (unspecified), Chlorothalonil, Iprodione, pyrethroids and Vinclozolin.

Residues found

The 1981–4 study: residues found below the MRLs have not been published.

Maximum Residue Limits

Tomatoes have 34 MRLs.

Wheat

Pesticides used

Approximately 90 are used during wheat production. In 1988 the most extensively used insecticides were Dimethoate, Demeton-s-methyl, Pirimicarb and Methiocarb; fungicides Carbendazim and Prochloraz; herbicides Isoproturon, Mecoprop, Ioxynil and Bromoxynil.

Extent of MAFF analysis

1. In a 1981 study 47 samples of fumigant residues in flour (bromomethane and carbon tetrachloride) were analysed.
2. During 1982 (A) 139 UK samples and 29 imported samples were analysed for organophosphates and fumigants (both groups unspecified).
3. During 1982 (B) seven samples of imported retail wheat products were analysed for Lindane and its impurities, and DDT and its breakdown products.
4. During 1984 182 samples of cereal products were analysed for an unspecified number of organophosphates.
5. Between 1984 and 1987 920 samples of retail wheat products were analysed for an unspecified number of organophosphates.
6. Between 1988 and 1989 58 samples of wheatgerm were analysed for 23 pesticides including Aldrin, Dieldrin, Endrin, Carbaryl, carbon tetrachloride, Chlorpyrifos-methyl, cis-Chlordane, DDT and its breakdown products, 1,2-Dibromoethane, Diazinon, Dichlorvos, Endosulfan (alpha- and beta-), Etrimfos, Fenitrothion, alpha-HCH, beta-HCH, gamma-HCH, Heptachlor, Hexachlorobenzene, Malathion, Methacrifos, Phosphamidon and Pirimiphos-methyl.
7. During 1988 120 UK bread samples were analysed for 20 pesticide residues including Aldrin, Chlordane, Chlorpyrifos-

methyl, DDT, Diazinon, Dichlorvos, Dieldrin, Endosulfan, Endrin, Etrimfos, Fenitrothion, Lindane and its impurities, Heptachlor, Hexachlorobenzene, Malathion, Methacrifos, Phosphamidon and Pirimiphos-methyl. In addition, 47 of the 120 samples were tested for an unspecified number of organochlorines.
8. During 1989 148 UK bread samples were analysed for 21 pesticides residues. They included those sought during 1988, plus Carbaryl.

Residues found

1. The 1981 study: 34% of samples contained residues of bromoethane and 21% contained residues of carbon tetrachloride.
2. The 1982 study (A): the overall percentages were not published. Residues of carbon tetrachloride, Chlorpyrifos-methyl, Pirimiphos-methyl and Malathion were found
3. The 1982 study (B): 43% of samples contained residues of DDT and its breakdown products and Lindane impurities.
4. The 1984 study: Pirimiphos-methyl was found in (29%) of wholemeal flour, (43%) of bran and (11%) of bran-based breakfast cereals.
5. Residues were detected of Chlorpyrifos-methyl, Etrimfos, Fenitrothion, Malathion and Pirimiphos-methyl.
6. The 1988–9 study: 72% of wheatgerm samples contained residues of Chlorpyrifos-methyl, Etrimfos, Fenitrothion, Malathion and Pirimiphos-Methyl.
7. The 1988 study: 32% of bread samples contained residues of Methacrifos, Malathion, Chlorpyrifos-methyl, Etrimfos, Fenitrothion, Lindane and its impurities and Pirimiphos-methyl (47% of wholemeal bread samples contained residues, whereas only 10% of white sliced bread samples contained residues).
8. The 1989 study: 22% of bread samples contained residues of Chlorpyrifos-methyl, Etrimphos, Lindane, Malathion, Pirimiphos-methyl (30% of non-wholemeal bread and 17% of white sliced samples contained residues).

Maximum Residue Limits

Wheat has 31 MRLs.

Table 2
The Main Pesticides Found

The following notes will help you understand Table 2. The letters and numbers denoting different categories also refer to Table 3. This list of pesticides is typical rather than comprehensive. Sixty have been chosen out of a total of approximately 440 cleared for use in the UK; they cover each of the major categories – insecticides, fungicides, herbicides, growth regulators and wood preservatives. Many of the most commonly used active ingredients are listed; some that are not in such widespread use, or that have been in the public eye (the apple growth regulator Daminozide, and the EBDC fungicides), have also been selected. Some pesticides that are not used in this country, but which may be found as residues in imported food (e.g. DDT), are included. If you do not find the pesticide you are looking up in Table 2 you should be able to find it in Table 3. In each case the common name of the pesticide, as opposed to the trade name or chemical name, is given.

Name

We use the common name in bold.

Type

The name of the chemical group to which the pesticide belongs – organochlorine, carbamate etc. – is given together with the principal use – insecticide, herbicide etc.

Uses

All pesticides are approved for quite specific uses, and it is illegal to use one for a purpose for which approval has not been given. The principal UK uses are summarised here, and major worldwide uses are also indicated where appropriate. Garden uses are also mentioned. Active ingredients that are approved for use in the UK are listed in *Pesticides 1990*; further details of uses are given on labels and in *The UK Pesticide Guide 1990* and *The Pesticide Users' Health and Safety Handbook* (see *Further Reading* on p. 235).

Acute or long term toxicity

a) Acute

Pesticide formulations are classified according to hazard by the World Health Organisation. The hazard referred to is the acute risk to health – the risk posed by single or multiple exposures over a relatively short time – that might be accidentally encountered by any person handling the product in accordance with the manufacturer's directions for handling; or in accordance with the rules laid down for storage and transportation by competent international bodies. The hazard categories are determined primarily by the acute toxicity of the pesticide formulation when eaten by rats or placed on their skin.

The classes are:

- IA – extremely hazardous
- IB – highly hazardous
- II – moderately hazardous
- III – slightly hazardous.

Other compounds are listed as 'unlikely to present acute hazard under normal use' (*WHO Recommended Classification of Pesticides by Hazard*, ref. VBC/86.1, Geneva, 1986–7). Here they are listed as 'Class 4' pesticides.

Other acute effects are quoted from information on labels or from *Recognition and Management of Pesticide Poisoning* (US EPA, 540/9–88–001, Washington DC, 1989) or from the references given.

b) Long-term effects IARC, EPA

There are two authoritative classifications of risk from cancer. The International Agency for Research on Cancer (IARC) publishes *Monographs on the Evaluation of Carcinogenic Risks to Humans* (Vols 1–42 and Supplement, IARC, Lyons, France,

1987). The US Environmental Protection Agency classifies cancer risk using similar listings. The schemes are as follows:

	IARC group	EPA group
● carcinogenic to humans	1	A
● probably carcinogenic to humans	2A	B1, B2
● possibly carcinogenic to humans	2B	C
● presently not classifiable as to their carcinogenicity to humans	3	D
● probably not carcinogenic to humans	4	E

Most pesticides have not been classified by either EPA or IARC. In some cases this is because they are not cancer risks; in other cases, the necessary research has simply not been done.

Other long-term possible effects on health are indicated as appropriate.

Regulatory status

This category includes pesticides under review by the Advisory Committee on Pesticides, which advises UK ministers; it also includes information about the actions of foreign governments in banning or severely restricting pesticides.

a) Bans and severe restrictions (SRs)

National regulatory authorities notify to the United Nations International Register of Potentially Toxic Chemicals (IRPTC) their regulatory actions concerning chemicals. Those actions which relate to pesticides – whether a pesticide has been banned from use by a country, or severely restricted in its use, are listed in this section from information provided by IRPTC and the publication *Consolidated List of Products Whose Consumption and/or Sale Have Been Banned, Withdrawn, Severely Restricted, or Not Approved by Governments* (UN, New York, 1989).

b) MAFF

A number of pesticides are being reviewed by MAFF. Details are shown of whether these pesticides are

● in the process of being reviewed (R)
● in the 'priority category' of pesticides for review (PR)
● to be reviewed at some future date (FR).

UK MRLs

- Where the UK has set its own Maximum Residue Levels, these are indicated by X. MRLs are the maximum residue levels allowed by law for a particular pesticide on food at the farm gate. MRLs are not safety levels. Unless otherwise stated, residues found in the following foods are found at levels below MRLs.

Water

Pesticides that fall within the Department of the Environment's list of substances dangerous to water (the 'Red List') are marked X. (Department of the Environment, *Dangerous Substances to Water*, 1988).

Environment

Where it says 'risk to bees' (or fish, cattle, etc.) this means there is a risk to that particular form of wildlife.

Effects

We have selected effects of particular significance. The list is only selective and not comprehensive. Specific references are given in the list of references on p. 252.

Alachlor

Type
Soil-acting anilide herbicide.

Uses
UK: pre-emergent use on oilseed rape, and on fodder maize to control broad-leaved and grass weeds; *worldwide* on maize and soyabeans.

Acute	Long-term toxicity		Regulatory status		UK MRLs	Water	Environment
WHO	IARC	EPA	Bans/SRs	MAFF			
III	–	B2	2	R	–	–	[Data gaps], fish

Effects

Mild irritant. Reported to be carcinogenic;[1] causes cancer in rats.[2] MAFF review requires 'further detailed reassurance regarding the absence of a genotoxic mechanism':[21]

Aldicarb

Type

Soil-applied systemic carbamate insecticide and nematicide.

Uses

UK: on vegetables (cabbages, cauliflowers and sprouts) and ornamental plants; *worldwide:* on sugar beet.

Acute	Long-term toxicity		Regulatory status		UK MRLs	Water	Environment
WHO	IARC	EPA	Bans/SRs	MAFF			
IA	–	–	8	FR	–	–	Water, bees

Effects

High oral toxicity; under review by EPA for reason of acute toxicity. Low concentrations in water found to affect human immune system.[3,4] The ACP disagrees. Found in ground water in USA below sandy soils, but not yet in UK – MAFF has asked UK water authorities to monitor.[5] Toxic to bees.[6] One of PAN 'Dirty Dozen' pesticides.

Amitraz

Type

Formamidine acaricide and insecticide.

Uses

UK: control of red spider mite in apples and pear sucker; lice control in animals.

Acute	Long-term toxicity		Regulatory status		UK MRLs	Water	Environment
WHO	IARC	EPA	Bans/SRs	MAFF			
III	–	C	2	FR	–	–	Birds

Effects

Some evidence of liver tumours in female mice.[7] Harmful to fish; concern over possible effects on bird reproduction and potential to contaminate water.[8]

Amitrole

Type

Residual translocated non-selective herbicide.

Uses

UK: broad-spectrum weed-killer used by local authorities, and on sweetcorn, wheat, kale, rape, potatoes, pears and apples; *garden* uses.

Acute	Long-term toxicity		Regulatory status		UK MRLs	Water	Environment
WHO	IARC	EPA	Bans/SRs	MAFF			
4	2B	B2	5	FR	X	–	Persistent in soil and water

Effects

Reported to cause cancer in mouse liver, and rat and mouse thyroid;[9] reviewed by MAFF in 1984[10] but not considered a human carcinogen provided consumption less than 2 parts per million. EPA has not permitted food uses or residues since 1971.

Atrazine

Type

Residual and foliar triazine herbicide.

Uses

UK: weed control in sweetcorn and raspberries; total weed control in conifer plantations, non-crop areas and field boundaries; *worldwide:* weeds in sweetcorn, fruit and vegetables; *garden* uses.

Acute	Long-term toxicity		Regulatory status		UK MRLs	Water	Environment
WHO	IARC	EPA	Bans/SRs	MAFF			
4	–	C	3	R	X	X	Leaches to water

Effects

Preliminary evidence of mammary tumours in female rats.[11] Weak mutagenic effect; some reproductive effects in rats.[12] Some UK local authorities propose to ban its use near water.[13]

Benomyl

Type
Systemic benzimidazole (MBC) fungicide.

Uses
UK: control of fungal diseases on wide variety of cereal crops, peas, beans, brassicas, other vegetables and fruit; approved for aerial application; *worldwide:* on fruit, vegetables, soyabeans, wheat and rice; *garden* uses.

Acute	Long-term toxicity		Regulatory status		UK MRLs	Water	Environment
WHO	IARC	EPA	Bans/SRs	MAFF			
4	–	C	3	–	–	–	–

Effects
Dermal sensitisation has occurred in agricultural workers exposed to foliage residues. Reproductive effects on rats have been observed with decreased sperm count at high dose levels. Embryotoxic and teratogenic effects on rats and mice at high dose levels. 'Weak',[20] mutagen at high dose levels.[19]

Bromoxynil

Type
Contact hydrobenzonitrile (HBN) herbicide used only in mixtures.

Uses
UK: controls broad-leaved post-emergent weeds in cereal crops.

Acute	Long-term toxicity		Regulatory status		UK MRLs	Water	Environment
WHO	IARC	EPA	Bans/SRs	MAFF			
II	–	–	2	R	–	–	Fish, water, bees

Effects
Developmental effects in rats: risks through potential operator exposure rather than food residues.[14] Concern about reproductive effects and effects on thyroid gland.[15] Following a review by ACP in 1985 and again in 1988, it was concluded that 'effects

on the thyroid and male reproductive capacity were not significant, but had greater concern about teratogenicity'. Approvals for home and garden use were revoked. 'Concern was also expressed . . . that the studies performed over the past 20 years had been poorly performed and the information provided was not complete or up to contemporary standards.'[16] Conditions of use varied.[17] Dangerous to fish; and because of scale of use of compound and detection in ground water, Dept of Environment was recommended to monitor pesticide in water.[18]

Captan

Type

Broad spectrum dicarboximide fungicide.

Uses

UK: control of scab in apples and pears, fungal diseases in strawberries and tomatoes; *garden* uses.

Acute	Long-term toxicity		Regulatory status		UK MRLs	Water	Environment
WHO	IARC	EPA	Bans/SRs	MAFF			
4	3	B2	3	R	–	–	Fish

Effects

Moderately irritating to skin, eyes and respiratory tract. Literature reports of possible mutagenicity[24] and carcinogenicity.[25,26] Harmful to fish: EPA concern over possible threat to endangered species.[27] EPA has banned uses on 42 different fruit and vegetables because of concerns over cancer risks in January 1989. Review by EPA because of concerns about oncogenicity and mutagenicity.

Carbaryl

Type

Contact carbamate insecticide

Uses

UK: insect control on apples, pears, soft fruit, brassicas, tomatoes; control of earthworms in turf and amenity grass; licensed as a veterinary pesticide to control fleas and lice; approved for some aerial applications; *worldwide:* on cotton; *garden* uses.

Acute	Long-term toxicity		Regulatory status		UK MRLs	Water	Environment
WHO	IARC	EPA	Bans/SRs	MAFF			
II	3	–	1	FR	X	–	Bees, fish

Effects

Anticholinesterase compound: workers should not use it if under medical advice not to work with such compounds. Acute exposure may cause nausea, muscle weakness, dizziness, sweating, blurred vision and muscle twitching. EPA class it as 'a weak mutagen';[25] a short-term study on humans demonstrated an apparent effect on kidney function.[22] Neurotoxic effects reported on exposed chickens, and rats.[23] Dangerous to bees and fish. Approved for some aerial applications.

Carbofuran

Type

Systemic carbamate insecticide and nematicide for soil treatment.

Uses

UK: wide range of uses on beets, brassicas, sweetcorn, potatoes and ornamentals; *worldwide* use on sweetcorn.

Acute	Long-term toxicity		Regulatory status		UK MRLs	Water	Environment
WHO	IARC	EPA	Bans/SRs	MAFF			
IB	–	–	2	FR	–	–	Fish, wildlife, water

Effects

Carboturan is an anti-chlolinesterase compound, and should not be used if under advice not to work with such compounds. Harmful in contact with skin. Adverse reproductive effects reported in rats.[28] Dangerous to livestock, fish, game, wild birds and animals. Subject to EPA Special Review because of concern over effects on wildlife: 40 bird death incidents reported.[29] Concern over potential to leach to ground water.[30]

Chlordane

Type

Persistent organochlorine insecticide.

Uses

UK: control of earthworms in turf; *worldwide:* to control termites, ants, household pests and pests on domestic animals.

Acute	Long-term toxicity		Regulatory status		UK MRLs	Water	Environment
WHO	IARC	EPA	Bans/SRs	MAFF			
II	3	C	32	–	–	–	Wildlife, fish, soil persistence

Effects

Harmful in contact with skin. Chlordane produces liver cancers in mice, but the results for rats are inconclusive.[31] Some evidence of mutagenicity.[32] Adverse reproductive effects reported at high doses in rats.[33] Harmful to fish; dangerous to livestock; highly toxic to earthworms; highly persistent. Bio-accumulation of chlordane less than other organochlorine pesticides: some evidence of small amounts of metabolite in human breastmilk.[34] One of the PAN 'Dirty Dozen'. UK use banned after 1992.

Chlorfenvinphos

Type

Soil-applied organophosphorous insecticide.

Uses

UK: insect control on winter wheat, root and leaf brassicas and sweetcorn; approved for some aerial applications.

Acute	Long-term toxicity		Regulatory status		UK MRLs	Water	Environment
WHO	IARC	EPA	Bans/SRs	MAFF			
IA	–	–	–	PR	X	–	Wildlife, fish

Effects

Chlorfenvinphos is an anti-cholinesterase agent and should not be used if advised not to work with organophosphorous compounds. Harmful to game, wild birds and animals; dangerous to fish.

Chlormequat

Type
Plant growth regulator.

Uses
UK: reduces stem growth on cereal crops and flowers; approved for some aerial applications.

Acute Long-term toxicity			Regulatory status		UK MRLs	Water	Environment
WHO	IARC	EPA	Bans/SRs	MAFF			
III	–	–	1	FR	–	–	

Effects
Harmful in contact with skin and if swallowed. Report of sows in Denmark producing smaller litters because of interference in oestrus after feeding on treated grain.[35] Not to be used on food crops. Approved for some aerial applications.

Chlorothalonil

Type
Organochlorine fungicide.

Uses
UK: wide range of uses on fruit, vegetables, winter wheat, turf and ornamentals; *worldwide:* on fruit, vegetables and soyabeans.

Acute Long-term toxicity			Regulatory status		UK MRLs	Water	Environment
WHO	IARC	EPA	Bans/SRs	MAFF			
4	3	B2	0	R	–	–	Fish, water

Effects
Irritation can be caused to skin, mucous membranes, eyes and respiratory tract. Evidence of kidney tumours in mice; results of study on rats awaited.[36] Harmful to fish.

Chlorpyrifos

Type
Organophosphorous insecticide.

Uses

UK: contact control of insects in cereal crops, fruit, vegetables, forestry, amenity grass and grasslands; approved for some aerial applications.

Acute	Long-term toxicity		Regulatory status		UK MRLs	Water	Environment
WHO	IARC	EPA	Bans/SRs	MAFF			
II	–	–	0	FR	–	–	Bees, fish

Effects

Organophosphorous compound: should not be used if under medical advice not to work with such compounds. Harmful in contact with skin and if swallowed; irritating to eyes and skin. Dangerous to bees; dangerous to fish.

Cypermethrin

Type

Contact synthetic pyrethroid insecticide.

Uses

UK: insect control in cereal crops, beans, oilseed rape, brassicas, apples, pears, soft fruit, ornamentals and flowers; *garden* uses.

Acute	Long-term toxicity		Regulatory status		UK MRLs	Water	Environment
WHO	IARC	EPA	Bans/SRs	MAFF			
II	–	C	–	R	–	–	Bees, fish

Effects

Irritating to eyes and skin; harmful in contact with latter, and may cause sensitisation. Benign lung cancers produced at highest dose in one species of mice only; neurotoxic effects on rats at high doses, and possibility of immune suppression.[37] Dangerous to bees; extremely dangerous to fish.

2,4-D

Type

Translocated chlorophenoxy herbicide.

Uses

UK: as post-emergent weed control in cereal crops and grass; in forestry and for control of aquatic weeds; *worldwide*: on maize and wheat; *garden* uses.

Acute	Long-term toxicity		Regulatory status		UK MRLs	Water	Environment
WHO	IARC	EPA	Bans/SRs	MAFF			
II	2B	D	0	R	–	–	Water, fish

Effects

Harmful to skin; irritating to eyes and mucous membranes. Occupational exposure may lead to headache, vomiting, mental confusion. Accidental poisonings leading to severe neuro-toxicity have been reported, and EPA considers 2,4-D may be teratogenic.[38] There has been considerable evidence to suggest an association between exposure to phenoxy herbicides and an increased incidence of soft-tissue sarcomas and malignant lymphomas.[39,40] Harmful to fish; keep livestock out of treated areas. Priority candidate for inclusion in DoE 'Red List' of substances dangerous to water. EPA notes significant data gaps, including residue chemistry, dermal toxicology, non-rodent chronic toxicity, teratogenicity (rabbit), mutagenicity and neurotoxicity.[41]

Daminozide

Type

Plant growth regulator.

Uses

UK: on apples, pears, flowers and ornamentals to ensure set of fruit, even ripening, and reduction of distance between nodes.

Acute	Long-term toxicity		Regulatory status		UK MRLs	Water	Environment
WHO	IARC	EPA	Bans/SRs	MAFF			
4	–	B2	Many	FR	–	–	Fish

Effects

Irritating to eyes. EPA was concerned about the 'inescapable direct correlation' between exposure to UDMH, a breakdown product of Daminozide, and 'the development of life-threatening tumours'[42] and initiated a Special Review. The formation of UDMH from Daminozide is accelerated by heat processing. A review by ACP in 1989 concluded that there was some evidence of questionable genotoxicity of UDMH in the literature, but that UDMH was not genotoxic. The ACP concluded that UDMH was carcinogenic in mice, but that safety levels were sufficiently large to protect consumers.[43] Harmful to fish. Sales of Daminozide for use on food crops were halted worldwide by the manufacturers in October 1989.

DDT

Type

Broad-spectrum organochlorine insecticide.

Uses

UK: none; *worldwide:* agriculture and public health.

Acute	Long-term toxicity		Regulatory status		UK MRLs	Water	Environment
WHO	IARC	EPA	Bans/SRs	MAFF			
II	2B	B2	45		–	X	Birds, bats

Effects

Can cause sensory disturbances, confusion and convulsions. Some evidence of reproductive and mutagenic effects in test animals.[44] DDT and metabolites bio-accumulate in fatty tissues; DDT breakdown products have caused eggshell thinning and breeding failures in birds of prey worldwide, the latter also in bats.[45] Residues of DDT found worldwide in food and human breastmilk.

Deltamethrin

Type

Contact and residual pyrethroid insecticide.

Uses

UK: insect control in a wide variety of cereal crops, peas, beans, brassicas and other vegetables, and fruit.

Acute	Long-term toxicity		Regulatory status		UK MRLs	Water	Environment
WHO	IARC	EPA	Bans/SRs	MAFF			
II	–	–	0	R	–	–	Bees, fish

Effects

Harmful in contact with skin; irritating to eyes and skin. Dangerous to bees; extremely dangerous to fish.

Demeton-s-methyl

Type

Systemic organophosphorous insecticide.

Uses

UK: control of aphids and mites on a wide range of fruit and vegetables; *worldwide:* on fruit and vegetables.

Acute	Long-term toxicity		Regulatory status		UK MRLs	Water	Environment
WHO	IARC	EPA	Bans/SRs	MAFF			
IB	–	–	1	R	–	–	Wildlife, bees, fish

Effects

Organophosphorous compound: not to be used if under medical advice not to work with such compounds. Toxic in contact with skin, by inhalation, and if swallowed. EPA cites many data gaps, including neurotoxicity, chronic toxicity, oncogenicity, teratogenicity, reproduction and environmental fate. Harmful to game, birds, wild animals, livestock and bees; dangerous to fish.

Diazinon

Type

Contact organophosphorous insecticide:

Uses

UK: control of soil insects in leaf brassicas, carrots and lettuce, tomato and cucumber pests; licensed veterinary medicine for sheep-dipping; *worldwide:* on rice.

Acute	Long-term toxicity		Regulatory status		UK MRLs	Water	Environment
WHO	IARC	EPA	Bans/SRs	MAFF			
II	–	–	0	R	–	–	Wildlife, fish

Effects

Diazinon is an organophosphorous compound: not to be used if under medical advice not to work with such compounds. Dermal sensitisation caused in 10% of volunteers tested.[46] EPA has identified data gaps for reproduction and mutagenicity studies.[47] Harmful to game, wild birds, animals and fish. EPA concern as to potential hazard for birds led to cancellation of uses on turf farms and golf courses in USA in 1987. Evidence of persistence in water and bio-accumulation in fatty tissue of snails.[48]

Dicamba

Type

Translocated benzoic herbicide.

Uses

UK: weed control in cereals, and control of bracken in forestry and non-crop areas; *garden* uses.

Acute	Long-term toxicity		Regulatory status		UK MRLs	Water	Environment
WHO	IARC	EPA	Bans/SRs	MAFF			
4	–	–	0	FR	–	–	–

Effects

Moderately irritating to skin and respiratory tract. Keep livestock out of treated area.

Dichlorvos

Type

Organophosphorous insecticide.

Uses

UK: insect control and fumigant in glasshouses; licensed veterinary pesticide to control fleas on animals and fish; *garden* uses.

Acute	Long-term toxicity		Regulatory status		UK MRLs	Water	Environment
WHO	IARC	EPA	Bans/SRs	MAFF			
IB	3	B2		R	X		X	Wildlife, fish water

Effects

Organophosphorous compound: not to be used if under medical advice not to work with such compounds. Immune suppression has been reported in rabbits.[49] Some evidence of carcinogenicity for male rats and mice; clear evidence of carcinogenicity for female mice and evidence of mutagenicity.[50] Dangerous to bees and fish; harmful to game, wild birds and animals. Concerns over use of Dichlorvos in fish farming.[51]

Dicofol

Type

Non-systemic organochlorine acaricide.

Uses

UK: control of red spider mites in glasshouses and on tomatoes, cucumbers, hops, apples and strawberries.

Acute	Long-term toxicity		Regulatory status		UK MRLs	Water	Environment
WHO	IARC	EPA	Bans/SRs	MAFF			
III	3	B2	4	FR	X	–	Fish, birds

Effects

Harmful in contact with skin; irritating to eyes and respiratory system. Limited evidence of carcinogenicity in mice.[52] Can cause reproductive impairment (eggshell thinning) in various fish and flesh-eating birds.[53] Highly toxic to aquatic organisms. Bio-accumulates in rotational crops and aquatic organisms.[54]

Dimethoate

Type

Broad-spectrum contact organophosphorous insecticide.

Uses

UK: used to control aphids and mites in a wide variety of fruit, vegetables and cereal crops; *worldwide:* fruit, vegetables, cotton and wheat; *garden* uses.

Acute	Long-term toxicity		Regulatory status		UK MRLs	Water	Environment
WHO	IARC	EPA	Bans/SRs	MAFF			
II	–	–	2	PR	X	–	Wildlife, fish

Effects

Organophosphorous compound: not to be used if under medical advice not to work with such compounds. Harmful in contact with skin. Foetotoxic in rats.[55] There is contradictory information available concerning the possible carcinogenicity of Dimethoate.[56] Harmful to livestock, game, wild birds, animals and fish; dangerous to bees.

Dinocap

Type

Dinitrophenol fungicide.

Uses

UK: control of powdery mildew in soft fruit, hops, cucumbers and flowers; *worldwide:* on sweetcorn.

Acute	Long-term toxicity		Regulatory status		UK MRLs	Water	Environment
WHO	IARC	EPA	Bans/SRs	MAFF			
III	–	–	1	FR	–	–	Fish

Effects

Harmful by inhalation and in contact with skin; irritating to eyes, skin and respiratory system. Variety of embryotoxic and teratogenic effects observed in rabbits, mice, rats and hamsters; oncogenic study required in second species.[57] Dangerous to fish. MAFF review in progress: home and garden approvals revoked pending further data.

Endosulfan

Type

Organochlorine insecticide.

Uses

UK: oilseed rape, berries, hops and narcissi; *worldwide:* rice.

Acute	Long-term toxicity		Regulatory status		UK MRLs	Water	Environment
WHO	IARC	EPA	Bans/SRs	MAFF			
II	–	–	13	FR	X	X	Bees, fish, livestock

Effects

Harmful in contact with skin. EPA considers Endosulfan teratogenic. Harmful to bees and livestock; dangerous to fish.

Formaldehyde

Type

Fungicide and soil sterilant.

Uses

UK: glasshouse fumigant.

Acute	Long-term toxicity		Regulatory status		UK MRLs	Water	Environment
WHO	IARC	EPA	Bans/SRs	MAFF			
–	2B	B2	0	R	–	–	Water

Effects

Harmful in contact with skin or by inhalation; irritant to eyes and upper respiratory tract; can sensitise, leading to asthma or dermatitis. Mutagenic effects reported.[58] EPA concerned about effects on water and aquatic species; data gaps identified.[59]

Glyphosate

Type

Translocated phosphonate herbicide.

Uses

UK: control of annual and perennial weeds in a wide variety of field crops, and in grassland, orchards and forestry; *garden* uses.

Acute	Long-term toxicity		Regulatory status		UK MRLs	Water	Environment
WHO	IARC	EPA	Bans/SRs	MAFF			
4	–	D	0	FR	–	–	Fish

Effects

Irritating to eyes, skin and upper respiratory tract. Risk of serious damage to eyes. Oncogenic potential of Glyphosate not fully understood.[60] Dangerous to fish.

Ioxynil

Type

Contact hydrobenzonitrile (HBN) herbicide.

Uses

UK: annual weeds in leeks, onions, garlic and turf.

Acute	Long-term toxicity		Regulatory status		UK MRLs	Water	Environment
WHO	IARC	EPA	Bans/SRs	MAFF			
II	–	–	0	R	X	–	Fish, water, bees

Effects

Irritating to eyes and skin; harmful in contact with skin. Keep livestock out of treated areas; harmful to bees and fish. Following a review by ACP in 1985, it was concluded that 'effects on the thyroid and male reproductive capacity were not significant, but had greater concern about teratogenicity.' Approvals for home and garden use were revoked. 'Concern was also expressed . . . that the studies performed over the past 20 years had been poorly performed and the information provided was not complete or up to contemporary standards.'[61] Conditions of use varied.[62] Because of scale of use, and detection in ground water, Dept of Environment was recommended to monitor in water.[63]

Iprodione

Type

Broad-spectrum dicarboximide fungicide.

Uses

UK: wide range of uses on cereals, vegetables, fruit and ornamentals; approved for some aerial applications.

Acute	Long-term toxicity		Regulatory status		UK MRLs	Water	Environment
WHO	IARC	EPA	Bans/SRs	MAFF			
4	–	–	0	FR	X	–	Fish

Effects

Irritating to eyes and skin. Harmful to fish. No results of long-term toxicity studies were available. ACP concerned about residue levels from multiple pre-harvest treatments of lettuce, but questioned whether action over residue levels was justified given that no evidence existed or adverse effects.[63]

Isoproturon

Type

Substituted urea herbicide.

Uses

UK: widely used in cereals; approved for some aerial applications.

Acute	Long-term toxicity		Regulatory status		UK MRLs	Water	Environment
WHO	IARC	EPA	Bans/SRs	MAFF			
III	–	–	0	PR	–	–	Water

Effects

Irritating to eyes, skin and mucous membranes. Leaches to water.

Lindane

Type

Persistent organochlorine insecticide.

Uses

UK: insect control in cereals crops, grassland, stored grain, fruit and forestry; insecticide in timber treatment products.

Acute	Long-term toxicity		Regulatory status		UK MRLs	Water	Environment
WHO	IARC	EPA	Bans/SRs	MAFF			
II	3	C	25	R	X	X	Fish, bees, water

Effects

Toxic or harmful in contact with skin, or by inhalation; irritating to eyes and respiratory system. Concern over possible association with aplastic anaemia.[64,65] Dangerous to bees and fish; harmful to livestock. Increasing levels of Lindane in seawater.[66] HSE review list. One of PAN 'Dirty Dozen'.

Malathion

Type

Broad-spectrum contact organophosphorous insecticide.

Uses

UK: control of aphids and other insect pests in a wide range of fruit, vegetables and flowers; licensed veterinary medicine for control of lice; approved for some aerial applications; *worldwide:* on rice; *garden* uses.

Acute	Long-term toxicity		Regulatory status		UK MRLs	Water	Environment
WHO	IARC	EPA	Bans/SRs	MAFF			
III	3	–	0	PR	X	X	Fish, bees, water

Effects

Organophosphorous compound: do not use if under medical advice not to work with such compounds. EPA identifies data gaps including teratogenicity and reproductive effects.[67] Harmful to bees and fish. EPA concern from reports of fish deaths and field studies that adverse effects to aquatic and land-based fauna may result from normal use.[68]

Mancozeb

Type

Fungicide of the ethylene bisdithiocarbamate type (EBDC).

Uses

UK: control of fungal disease in cereals, oilseed rape, fruit and flowers; approved for some aerial applications; *garden* uses.

Acute	Long-term toxicity		Regulatory status		UK MRLs	Water	Environment
WHO	IARC	EPA	Bans/SRs	MAFF			
4	–	B2	0	R	–	–	Fish

Effects

Irritating to respiratory system; may cause sensitisation by skin contact. No test for this fungicide or its breakdown product in body fluids is available. Ethylene thiourea (ETU), a common contaminant, metabolite and degradation production of EBDC, is classed as a carcinogen. IARC regards ETU as a Class 2B carcinogen (possibly carcinogenic to humans: inadequate human evidence, sufficient animal data). EPA classes ETU as a B2 carcinogen (probable human carcinogen), and considers there are carcinogenic risks to consumers from dietary exposure to ETU, and carcinogenic, developmental and thyroid risks to mixers, loaders and applicators.[69] In its review, ACP considered there were no risks to consumers of thyroid tumours, or developmental effects; the significance of rodent tumours to humans was considered uncertain.[70] Harmful to fish. EPA announced its intention in 1998 to cancel 40 food uses of EBDCs, and manufacturers agreed voluntarily to withdraw the products from use.[71] EBDCs have not been withdrawn in UK.

Maneb

Type

Fungicide of the ethylene bisdithiocarbamate type (EBDC).

Uses

UK: control of fungal disease in potatoes, cereal crops, brassicas, sugar beet and tomatoes; approved from some aerial applications.

Acute	Long-term toxicity		Regulatory status		UK MRLs	Water	Environment
WHO	IARC	EPA	Bans/SRs	MAFF			
4	3	B2	0	R	–	–	Fish

Effects

Harmful if swallowed; irritating to skin, eyes and respiratory system; may cause sensitisation by skin contact. No test for this fungicide or its breakdown products in body fluid is available. Ethylene thiourea (ETU) is a common contaminant, metabolite and degradation product of EBDCs. IARC classes ETU as a Class 2B carcinogen (possible human carcinogen: inadequate human data, sufficient animal data). EPA classes

ETU as a B2 carcinogen (probable human carcinogen). EPA considers there are carcinogenic risks to consumers from dietary exposure to ETU, and carcinogenic, developmental and thyroid risks to mixers, loaders, and applicators.[72] IARC classes Maneb as a Class 3 carcinogen (not classifiable as to human carcinogenicity: inadequate animal data, no human data). In its review, ACP considered there were no risks to consumers from Maneb or ETU of thyroid tumours or developmental effects; the significance of rodent tumours to humans was considered uncertain.[73] Harmful to fish and domestic fowl.[74] EPA announced its intention in 1989 to cancel 40 food uses of EBDCs, and manufacturers agreed voluntarily to withdraw the products from use.[75] EBDCs have not been withdrawn in UK.

MCPA

Type
Translocated phenoxy acid herbicide.

Uses
UK: annual weeds in cereals, asparagus, grassland, turf and amenity areas; *garden* uses.

Acute Long-term toxicity			Regulatory status		UK MRLs	Water	Environment
WHO	IARC	EPA	Bans/SRs	MAFF			
III	2B	B2	1	PR	–	–	Water

Effects
Harmful if swallowed; harmful in contact with skin; risk of serious danger to eyes; EPA has identified data gaps requiring additional studies in teratology, reproduction, mutagenicity and neurotoxicity.[76] Keep livestock out of treated areas. EPA concern over potential to contaminate ground water.[77]

Mecoprop

Type
Translocated phenoxy herbicide.

Uses

UK: control of post-emergent weeds in cereals, grassland and orchards; *worldwide:* on wheat; *garden* uses.

Acute	Long-term toxicity		Regulatory status		UK MRLs	Water	Environment
WHO	IARC	EPA	Bans/SRs	MAFF			
III	2B	–	0	PR	–	–	Water

Effects

Harmful in contact with skin and if swallowed; irritating to eyes and skin. Concern over possible Dioxin contaminants in manufacture; EPA has identified data gaps concerning teratogenicity and dermal exposure.[78] Keep livestock away from treated areas. EPA concern over potential ground water contamination.

Methyl bromide

Type

Soil fumigant; only available in combination with other active ingredients.

Acute	Long-term toxicity		Regulatory status		UK MRLs	Water	Environment
WHO	IARC	EPA	Bans/SRs	MAFF			
–	3	–	8	R	–	–	Wildlife, bees, fish, ozone

Effects

Very toxic by inhalation; can cause burns; risk of serious damage to eyes; danger of serious damage to health from prolonged exposure; central nervous system depressant, may cause convulsions. Many data gaps identified by EPA, particularly concerning long-term effects.[79] Dangerous to game, wild birds, animals, bees and fish. Suspected ozone-depleter.[80]

Paraquat

Type

Non-selective, contact quaternary bipyridilium herbicide.

Uses

UK: weed control in a wide range of cereal crops, beans, beets, potatoes and fruit, and in forestry; *garden* uses.

Acute	Long-term toxicity		Regulatory status		UK MRLs	Water	Environment
WHO	IARC	EPA	Bans/SRs	MAFF			
II	–	E	5	PR	–	–	Animals

Effects

Toxic if swallowed – may be fatal; irritating and harmful in contact with skin; irritating also to eyes. There is no effective antidote. Fatal poisonings are reported to have occurred as a result of protracted dermal contamination by Paraquat. When ingested in sufficient dosage, Paraquat has life-threatening effects on the gastro-intestinal tract, kidney, liver, heart and other organs. It is implicated in many suicides and para-suicides, especially in Third World. Carcinogen in rats in high doses, and 'weakly genotoxic'.[81] Dangerous to livestock and wild animals.

Pentachlorophenol (PCP)

Type

Chlorinated phenol fungicide.

Uses

UK: fungicide in timber treatment.

Acute	Long-term toxicity		Regulatory status		UK MRLs	Water	Environment
WHO	IARC	EPA	Bans/SRs	MAFF			
IB	3	C	16	FR	–	X	Fish, air

Effects

Irritating to eyes, skin and respiratory system; harmful in contact with skin and if swallowed; contact dermatitis occurs commonly in workers having contact with PCP. Carcinogenic in mice;[82] cases of aplastic anaemia, peripheral neuropathy and leukaemia have been reported which were associated temporally with PCP exposure: causal relationships in these cases were not established.[83] Foetotoxic in rats; immunotoxic in rats, mice, chickens and cattle.[84] Epidemiological evidence links occupational exposure to mixtures of chlorophenols with increases of soft tissue sarcomas, nasopharyngeal cancers and lymphomas.[85] Dangerous to fish; harmful to livestock. High volatility can lead to PCP in air, indoors and outdoors, following timber treatment. High potential for bio-accumulation.

Permethrin

Type

Broad-spectrum pyrethroid contact insecticide.

Uses

UK: insect control on fruit, salad vegetables, flowers and ornamentals; *garden* uses.

Acute Long-term toxicity			Regulatory status		UK MRLs	Water	Environment
WHO	IARC	EPA	Bans/SRs	MAFF			
II	–	–	0	R	–	–	Fish

Effects

Irritating to eyes, skin and respiratory system. Dangerous to bees and fish.

Phenylmercury acetate

Type

Organomercury fungicide.

Uses

UK: cereal seed treatment.

Acute Long-term toxicity			Regulatory status		UK MRLs	Water	Environment
WHO	IARC	EPA	Bans/SRs	MAFF			
IA	–	–	11	–	–	X	Fish, birds

Effects

Harmful to skin, by inhalation, or if swallowed; can cause skin rashes. 'The mercurial fungicides are among the most toxic pesticides ever developed, in terms of chronic as well as acute hazard. Epidemics of severe, often fatal, neurologic disease have occurred when indigent residents of less developed countries consumed methyl mercury treated grain intended for the planting of crops.'[86] Harmful to fish, and to fish-eating and grain-eating birds.[87]

Prochloraz

Type

Broad-spectrum systemic imidazole fungicide.

Uses

UK: control of fungal disease in cereals, oilseed rape and ornamentals.

Acute	Long-term toxicity		Regulatory status		UK MRLs	Water	Environment
WHO	IARC	EPA	Bans/SRs	MAFF			
III	–	C	0	FR	–	–	Fish

Effects

Irritating to eyes and skin. Dangerous to fish.

Propiconazole

Type

Systemic triazole fungicide.

Uses

UK: cereals and oilseed rape; approved for some aerial applications; *garden* uses.

Acute	Long-term toxicity		Regulatory status		UK MRLs	Water	Environment
WHO	IARC	EPA	Bans/SRs	MAFF			
II	–	C	0	FR	–	–	Fish

Effects

Irritating to eyes, skin and respiratory system. Dangerous to fish.

Simazine

Type

Persistent soil-acting triazine herbicide.

Uses

UK: control of annual weeds in beans, fruit, sweetcorn, forestry and non-crop amenity areas; *worldwide:* on fruit and vegetables; *garden* uses.

Acute	Long-term toxicity		Regulatory status		UK MRLs	Water	Environment
WHO	IARC	EPA	Bans/SRs	MAFF			
4	–	–	0	R	–	X	Fish, water

Effects

Irritating to eyes, respiratory system and skin. EPA has identified data gaps including oncogenicity, teratogenicity and mutagenicity.[88] May be toxic to fish. Some UK local authorities have proposed severe restrictions on use because of concerns over water.[89]

Tecnazene

Type

Fungicide and growth regulator.

Uses

UK: to control fungal disease in tomatoes, lettuce and flowers; potato sprout suppressant; *garden* uses.

Acute	Long-term toxicity		Regulatory status		UK MRLs	Water	Environment
WHO	IARC	EPA	Bans/SRs	MAFF			
4	–	–	0	R	X	–	Fish, water

Effects

Irritating to eyes and respiratory system. WHO recommend that the genotoxicity and cell transformation capacity of Tecnazene should be studied.[91] ACP, when reviewing Tecnazene, expressed concern at the lack of mutagenicity data, and requested these gaps be filled 'as quickly as possible'.[92] ACP considers that Tecnazene has no tumourogenic potential.[93] Harmful to fish; DoE has expressed concern about effect on fisheries and marine and freshwater environment of discharges containing Tecnazene from potato washing plants.[94]

Thiabendazole

Type
Systemic MBC fungicide.

Uses
UK: on potatoes, oilseed rape, asparagus and mushrooms.

Acute	Long-term toxicity		Regulatory status		UK MRLs	Water	Environment
WHO	IARC	EPA	Bans/SRs	MAFF			
4	–	–	0	R	X	–	Fish

Effects
Harmful to fish.

Triadimefon

Type
Systemic triazole fungicide.

Uses
UK: widely used on cereals, grapes and vegetables; approved for certain aerial applications; *worldwide:* wheat.

Acute	Long-term toxicity		Regulatory status		UK MRLs	Water	Environment
WHO	IARC	EPA	Bans/SRs	MAFF			
III	–	–	0	FR	–	–	Fish

Effects
Irritating to eyes. Harmful to livestock and fish.

Tributyltin oxide (TBTO)

Type
Organotin fungicide.

Uses
UK: wood preservative and marine anti-fouling biocide.

Acute	Long-term toxicity		Regulatory status		UK MRLs	Water	Environment
WHO	IARC	EPA	Bans/SRs	MAFF			
IB	–	–	0	FR	–	X	Fish

Effects

Irritating to eyes, skin and respiratory system. EPA considers TBTO to be teratogenic. ACP reviewed TBTO and considered it may be teratogenic in rats and mice; it is recommending the revocation of a number of approvals.[90] Dangerous to fish.

Trichlorfon

Type

Contact organophosphorous insecticide.

Uses

UK: control of insect pests of beets, brassicas, fruit and celery; approved for some aerial applications.

Acute	Long-term toxicity		Regulatory status		UK MRLs	Water	Environment
WHO	IARC	EPA	Bans/SRs	MAFF			
III	3	–	0	FR	–	–	Fish

Effects

Organophosphorous compounds – not to be used if under medical advice not to work with such compounds. EPA classes as a possible teratogen – some effects noted in rats and pigs at high doses;[95] some evidence of immunological effects noted;[96] delayed neuropathy has been reported following human poisoning.[97] Harmful to fish.

Trifluralin

Type

Dinitroaniline soil-acting herbicide.

Uses

UK: annual weed control in winter cereals, beets, beans, fruit and brassicas; *worldwide:* cotton and soyabeans.

Acute	Long-term toxicity		Regulatory status		UK MRLs	Water	Environment
WHO	IARC	EPA	Bans/SRs	MAFF			
4	–	C	3	FR	–	X	Fish

Effects

Irritating to eyes, skin and respiratory system. Dangerous to fish.

Zineb

Type

Fungicide of the ethylene bisdithiocarbamate type.

Uses

UK: control of fungal disease in fruit, flowers, hops, potatoes, lettuce and tomatoes; approved for aerial application.

Acute	Long-term toxicity		Regulatory status		UK MRLs	Water	Environment
WHO	IARC	EPA	Bans/SRs	MAFF			
4	3	B2	–	R	–	–	Fish

Effects

Irritating to eyes, skin and respiratory system. No test for this fungicide or its breakdown products in body fluids is available. Ethylene thiourea (ETU) is a common contaminant, metabolite and degradation product of EBDCs. IARC classes ETU as a Class 2B carcinogen (possible human carcinogen: sufficient animal data, inadequate human data). EPA classes ETU as a B2 carcinogen (probable human carcinogen). EPA considers there are carcinogenic risks to consumers from dietary exposure to ETU, and carcinogenic, developmental and thyroid risks to mixers, loaders and applicators.[98] In its review, ACP considered there were no risks to consumers from Zineb or ETU of thyroid tumours or developmental effects; the significance of rodent tumours to humans was considered uncertain.[99] Adverse reproductive effects in women reported.[100] Harmful to fish and chickens. EPA announced its intention in 1998 to cancel 40 food uses of EBDCs, and manufacturers agreed voluntarily to withdraw the products from use.[101] EBDCs have not been withdrawn in UK.

Table 3
The Complete UK Approved List

This is the MAFF pesticides list taken from *Pesticides 1990: Pesticides Approved under the Control of Pesticides Regulations 1986*, Reference Book 500, Ministry of Agriculture, Fisheries and Food, Health and Safety Executive. Total number = 356.

For explanations of the abbreviations used in this table, see pages 182–185.

Pesticide	WHO	IARC	EPA	WATER	MAFF
Alachlor	III		B2		R
Aldicarb	IA				FR
Alkylaryl trimethyl ammonium chloride	–				
Allethrin	III				
Alloxydim-sodium	–				
Alphachloralose	–				
Alphacypermethrin	II				
Aluminium ammonium sulphate	–				
Aluminium phosphide	F				
Aluminium sulphate	–				FR
2-Aminobutane	–				R
Amitraz	III		C		FR
Amitrole	4	2B	B2		
Ammonium carbonate	–				FR
Ammonium sulphamate	4				FR
Ammonium sulphate	4				FR
Anthraquinone	4				
Asulam	4		C		FR
Atrazine	4		C	X	R
Azamethiphos	III				
Azinphos-methyl	IB		C	X	FR
Aziprotryne	4				FR
Bacillus thuringiensis	–				FR
Barban	III				
Benazolin	4				FR
Bendiocarb	II				FR
Benfuracarb	IB				
Benodanil	4				FR
Benomyl	4		C		
Bentazone	III				FR
Benzalkonium chloride	–				FR
Benzethonium-chloride	–				

Pesticide	WHO	IARC	EPA	WATER	MAFF
Benzoylprop-ethyl	III				FR
Bifenox	4				FR
Bitenthrin	II		C		
Bioallethrin	II				FR
Bioresmethrin	4				R
Bitumen	–		2B		FR
Bone oil	–				
Boric acid	4				
Brodifacoum	IA				
Bromacil	4				
Bromadiolone	IA				FR
Bromophos	III				FR
Bromoxynil	II				R
Bupirimate	4				FR
Butoxycarboxim	IB				R
Calciferol	–				FR
Captan	4	3	B2		R
Carbaryl	II	3			FR
Carbendazim	4				R
Carbetamide	4				FR
Carbofuran	IB				FR
Carbophenothion	IB				
Carbosulfan	II				
Carboxim	4				FR
Cetrimide	–				FR
Chloramben	4				FR
Chlordane	II	3	C		
Chlorfenvinphos	IA				PR
Chloridazon	4				FR
Chlormequat	III				FR
Chlormequat chloride	–				
2-Chloroethylphosphonic acid	–				FR
Chlorophacinone	IA				FR
Chloropicrin	–				FR
Chlorothalonil	4	3	B2		R
Chlorotoluron	4				FR
Chloroxuron	4				FR
Chlorophonium	4				FR
Chlorpropham	4	3			
Chlorpyrifos	II				FR
Chlorpyrifos-methyl	4				
Chlorsulfuron	4				
Chlorthal dimethyl	4				
Chlorthiamid	III				
Clofentezine	4				
Clopyralid	4				
Copper	–				
Copper ammonium carbonate	–				FR

Pesticide	WHO	IARC	EPA	WATER	MAFF
Copper complex – bordeaux mixture	–				FR
Copper hydroxide	–				FR
Copper oxychloride	4				FR
Copper sulphate	II				FR
Coumatetralyl	IB				FR
Cresylic acid	–				FR
Cufraneb	–				
Cyanazine	II				FR
Cyfluthrin	II				
Cymoxanil	III				FR
Cypermethrin	II		–		R
2,4-D	II	2B	D		R
Dalapon	4				FR
Daminozide	4	B2			FR
Dazomet	III				
2,4-DB	III				PR
Deltamethrin	4				R
Demeton-s-methyl	IB				R
Demeton-s-methyl sulphone	IB				R
2,4-DES	II				
Desmetryne	III				FR
Di-allate	II	3			
Diazinon	II				R
Dicamba	4				FR
Dichlobenil	4				FR
Dichlofluanid	4				FR
Dichlorophen	III				FR
1,2-Dichloropropane	F				
1,3-Dichloropropane	F		B2		
Dichlorprop	III				PR
Dichlorvos	IB	3	B2	X	R
Diclofop-methyl	III				FR
Dicloran	4				
Dicofol	III	3	B2		FR
Dienochlor	4				
Difenacoum	IA				FR
Difenzoquat	II				
Diflubenzuron	4				FR
Diflufenican	4				
Dikegulac	4				
Dimethoate	II				PR
Dinocap	III				R
Diphacinone	IA				FR
Diphenamid	III				
Diquat	II				PR
Disulfoton	IA				FR
Ditalimfos	4				
Dithianon	III				FR

Pesticide	WHO	IARC	EPA	WATER	MAFF
Dithiocarbamate complex	–				FR
Diuron	4				FR
DNOC	IB				R
Dodecylbenzyl trimethyl ammonium chloride	–				FR
Dodemorph	4				FR
Dodine	III				FR
Drazoxolon	II				FR
Endosulfan	II			X	FR
EPTC	II				
Ethiofencarb	II				
Ethirimol	4				FR
Ethofumesate	4				FR
Ethoprophos	IA				
Ethylmercury phosphate	–				
Etridiazole	III				FR
Etrimfos	II				FR
Fatty acids	–				FR
Fenarimol	4		D/E		
Fenbutin oxide	4				
Fenchlorphos	II				
Fenitrothion	II			X	FR
Fenoprop	III				FR
Fenpropidin	III				
Fenpropimorph	4				
Fentin acetate	II				PR
Fentin hydroxide	II				PR
Fenuron	4				FR
Fenvalerate	II				
Ferbam	4	3			R
Ferric sulphate	–				FR
Ferrous sulphate	–				FR
Flamprop-m-isopropyl	4				FR
Flamprop-methyl	III				
Flocoumafen	–				
Fluazifop-butyl	4				
Fluazifop-p-butyl	4				
Fluoroacetamide	IB				
Fluroxypyr	4				
Flusilazole	–				
Flutriafol	III				
Fonofos	IA				
Formaldehyde	F	2B	B2		FR
Formothion	II				PR
Fosamine-ammonium	4				FR
Fosetyl-aluminium	4				
Fuberidazole	III				FR
Furalaxyl	III				
Gibberellins	–				

Pesticide	WHO	IARC	EPA	WATER	MAFF
Glyphosate	4		D		FR
Grease	–				
Heptenophos	IB				FR
Hexazinone	III				FR
Hydrogen iodide	–				FR
8-Hydroxyquinoline	–				
Hymexazol	4				
Imazalil	–				FR
Imazamethabenze methyl	4				
Imazapyr	4				
Indol-3-ylacetic acid	–				FR
4-Indol-3-ylbutyric acid	–				
Iodofenphos	4				FR
Ioxynil	II				R
Iprodione	4				FR
Isoproturon	III				R
Isoxaben	4				
Lenacil	4				
Lindane	II	3	C	X	R
Linuron	4		C		R
Magnesium phosphide	F				
Malathion	III	3		X	PR
Maleic hydrazide	4	3			FR
Mancozeb	4		B2		R
Maneb	4	3	B2		FR
Manganese	–				
Manganese zinc dithiocarbamate	–				
Manganese zinc ethylene bisdithlocarbamate	–				
MCPA	III	2B	B2		PR
MCPB	III				PR
Mecoprop	III	2B			PR
Mecoprop-p	III				PR
Mefluidide	III				
Mephosfolan	IA				
Mepiquat	III				FR
Mercuric oxide	IB				
Mercurous chloride	II				
Metalaxyl	III				FR
Metaldehyde	III				PR
Metamitron	4				FR
Metazachlor	4				
Methabenzthiazuron	4				FR
Methacrifos	II				
Metham-sodium	II				
Methiocarb	II				PR
Methomyl	II				FR
Methoprene	IB				FR
2-Methoxyethylmercury acetate	–				

Pesticide	WHO	IARC	EPA	WATER	MAFF
Methyl bromide	F	3			R
Methyl nonyl ketone	–				
Metoxuron	4				
Metribuzin	4				FR
Metsulfuron-methyl	4				
Mevinphos	IA				
Mineral oil	–				
Monolinuron	4				R
Myclobutanil	III				
Nabam	II				R
Naled	II				FR
Naphthalene	–				FR
1-Naphthylacetamide	4				FR
1-Napthylacetic acid	4				FR
2-Napthyloxyacetic acid	–				
Napromide	4				FR
Nicotine	IB				PR
Nitrothal-isopropyl	4				
Nonylphenoxypoly (ethyleneoxy) ethanol-iodine complex	–				FR
Norbormide	II				FR
Nuarimol	II				
Octhilinone	III				FR
Ofurace	4				
Omethoate	IB				PR
Oryzalin	4		C		
Oxadiazon	4		B2		
Oxadixyl	III				
Oxamyl	III				
Oxine-copper	4				
Oxycarboxin	4				FR
Oxydemeton-methyl	IB				R
Paraformaldehyde	–				
Paraquat	II		E		PR
Penconazole	4				
Pencycuron	4				
Pendimethalin	III				FR
Pentachlorophenol	IB	3	C	X	FR
Pentanochlor	4				
Pepper	–				FR
Permethrin	II				R
Phenmedipham	4				
Phenols	–				FR
Phenothrin	4				
Phenylmercury acetate	IA				
Phorate	IA				PR
Phosalone	II				
Phoxim	II				FR

Pesticide	WHO	IARC	EPA	WATER	MAFF
Picloram	4				
Pine sawfly NVP	–				
Pirimicarb	II				PR
Pirimiphos-ethyl	IB				PR
Pirimiphos-methyl	III				PR
Potassium hydroxyquinoline	–				FR
Potassium sorbate	–				
Prochloraz	III		C		FR
Prometryn	4				FR
Propachlor	III				FR
Propamocarb hydrochloride	4				FR
Propham	4	3	C		FR
Propiconazole	II		C		FR
Propineb	4				R
Propoxur	II		B2		
Propyzamide	4				FR
Pyrazophos	II				FR
Pyrethrins	II				R
Pyridate	–				FR
Quassia	–				FR
Quinalphos	II				
Quinomethionate	4				FR
Quinonamid	4				
Quintozene	4	3			
Quizalofop-ethyl	III				
Reserpine	–				FR
Resmethrin	III				R
Rotenone	II				FR
Sethoxydim	II				FR
Simazine	4			X	R
Sodium carbonate	–				
Sodium chlorate	III				FR
Sodium cyanide	IB				PR
Sodium hydrogen carbonate	–				FR
Sodium metabisulphate	–				
Sodium monochloracetate	–				
Sodium propionate	–				
Sodium silver thiosulphate	–				
Sodium tetraborate	4				FR
Sulphur	4				FR
2,4,5-T	II	3	C		
Tar acids	–				
Tar oils	–				
2,3,6-TBA	III				FR
TCA	4				FR
Tebutam	4				
Tebuthiuron	III				FR
Tecnazene	4				R

Pesticide	WHO	IARC	EPA	WATER	MAFF
Terbacil	4				FR
Terbuthylazine	4				
Terbutryn	4		C		
Tetrachlorvinphos	4	3			FR
Tetradifon	4				FR
Tetramethrin	4				FR
Thiabendazole	4				R
Thifensulfuron methyl					
Thiofanox	IB				FR
Thiometon	IB				FR
Thionazin	IA				
Thiophanate-methyl	4				R
Thiourea	–				
Thiram	III	3			R
Tolclofos-methyl	4				
Tri-allate	III				FR
Triadimefon	III				FR
Triadimenol	III		C		FR
Triazophos	IB				PR
Trichlorfon	III	3			FR
Trichoderma viride	–				FR
Triclopyr	III				
(Z)-9-Tricosene	–				FR
Tridemorph	II				FR
Trietazine	4				FR
Trifluralin	4		C	X	FR
Triforine	4				FR
Vamidothion	IB				FR
Verticillium lecanil	–				FR
Vinclozolin	4				R
Warfarin	IB				FR
Wax	–				FR
Zinc	–				FR
Zinc oxide	–				FR
Zinc phosphide	IB				PR
Zineb	4	3	B2		R
Zineb-ethylene thiuram disulphide adduct	–				R
Ziram	III	3			

Table 4
The Symptoms of Pesticide Poisoning

These are just some of the symptoms that people have suffered from exposure to pesticides. The symptoms listed here are only illustrations and can, of course, be caused by other agents. Remember, too, that individual reactions and sensitivities can vary.

This is a summary of the symptoms from poisoning by the major groups of pesticides. We give only the adverse effects from relatively short-term exposure.[1]

For individual pesticides, especially for their long-term effects, you can look up each commonly used pesticide in Table 3. In the first paragraph of each pesticide listed there you will find which group any particular pesticide belongs to. For example, under 'Chlorfenvinphos' (p. 191) the second word is organophosphorus, which you will find below.

Table 4 gives the complete list of officially approved pesticides. If you have any doubts go to your doctor and read the information on how to complain (Chapter 14). The British Medical Association has asked doctors to take more interest in pesticide poisoning, so you can expect a sympathetic ear.[2]

Carbamates

These substances act similarly to organophosphates (see below), but with one difference. The action of carbamates is reversible: you can get an antidote. The symptoms of carbamate poisoning are similar to those of organophosphates, except that they act more quickly and recovery is also quicker. This, it is claimed by manufacturers and some experts, makes them safer as you become incapacitated before you get a bad dose. Skin allergies and red papules (little spots) on the skin can occur.

Chlorophenoxy compounds (2,4-D, Mecroprop)

These can be absorbed through the skin. Inhalation may cause skin, nose, eye and throat irritation or burning. Dizziness and unsteady walk can occur. Other pains may be felt in the chest and guts, and there may be muscle tenderness. If you swallow chlorophenoxy compounds the symptoms are as if you had inhaled them, plus mental confusion and speech difficulties.

Dipyridyls (e.g. Paraquat)

These affect the mucous membranes of your lungs and gut. Watch out for nose bleeds. Other symptoms are breakdown of the cornea, nose, skin or fingernails. The damage to the skin and lungs is delayed for a few days after exposure. The lungs selectively accumulate Paraquat, so the lung cells accumulate toxic levels. More serious poisonings are indicated by muscle ache and pain, leading to convulsions and possible death.

Organochlorines

To understand this group's symptoms, it helps to know how organochlorines work. They combine with the energy-producing enzyme system at nerve endings in the central nervous system. The exception is DDT, now banned in the UK, but still showing up as residues. DDT affects the peripheral nerves.

The symptoms of organochlorine poisoning show as 'nervous' action. Symptoms include the tremors and convulsions characteristic of DDT poisoning, and in addition dizziness, hyper-excitability, sensitivity to noise, headaches, restlessness and twitching eyelids, frequent among other organochlorine poisonings. Organochlorines are also liver poisons, some being converted to poisons of greater toxicity.

A typical pattern of symptoms from long-term low exposure to organochlorines is that of 40 people with confirmed exposure to Lindane/Pentachlorophenol. All symptoms were confirmed by doctors. Nearly all complained of constant weakness and dizziness, while more than half complained of abdominal pains or of an acne-like skin rash. Several people had additional psychosomatic complaints – aggressiveness or restlessness. Nearly half complained of sickness, and more than a third had a fast heartbeat.[3]

Organophosphates

These pesticides affect the nerve junctions. To fire a nerve impulse across a junction requires the body to produce a substance called acetylcholine. This must be broken down immediately after use by another chemical, an enzyme called acetylcholinesterase. Organophosphates prevent this substance working properly, by locking on to the enzyme. Hence the organophosphate pesticide is called an inhibitor. The locking-on is a two-stage process: the first is reversible, but the second – called ageing – is not. So the nerve keeps firing.

The usual symptoms of mild poisoning are everyday headache, fatigue and mild indigestion. Other symptoms are tremors, dizziness and twitching. Flu-like symptoms can be associated with organophosphate poisoning. Some of the chemicals (e.g. Dimethoate) produce an anaesthetic effect before the inhibition of cholinesterase.

More serious poisoning produces cramps, diarrhoea, vomiting, lung failure and, in the worst cases, convulsions, heartblock and death.

Phthalimides (e.g. Captan, Folpet)

Acute exposure can cause diarrhoea, weight loss, vomiting and skin, eye and throat irritation. These substances are nerve toxins and can also affect the breakdown of glucose sugar in the body.

Pyrethroids

These are nerve poisons. They interact with the sodium channels in membranes causing a train of impulses rather than a single impulse. Look out for dizziness and fatigue, facial twitching, burning sensations, tightness and numbness.

Triazines and ureas (e.g. Atrazine, Monuron)

These affect the mucous membranes of the skin, lungs and eyes. Symptoms reflect irritation of these areas.

Appendix 1
The European Consumers' Pesticide Charter

The European Consumers' Pesticide Charter was first published on 31 May 1990. It was launched simultaneously throughout Europe and its signatories included:

- *United Kingdom*
Parents for Safe Food
The Pesticides Trust
The Soil Association
The Women's Environmental Network

- *Germany*
Die Verbraucher Initiative
Pestizid Aktions-Netzwerk

- *Switzerland*
Aktion Gesunder Essen

- *Denmark*
NOAH

- *Europe-wide*
Pesticides Action Network (PAN) Europe (based in Holland)

- *Italy*
Agrisalus
Movimento dei Consumatori Veneto

- *Luxembourg*
Mouvement Ecologique

- *Norway*
Norges Naturvernforbund

- *outside Europe*
Pesticides Action Network (PAN) Asia Pacific

For details contact Parents for Safe Food or the Pesticides Trust (addresses on p. 234).

General approach

There is growing concern among consumers worldwide about the impact of agrochemicals in the food chain. We, the consumers, can help improve both our own and the environment's health by using our influence to reduce agrochemical usage. This European Consumers' Pesticide Charter outlines some key improvements we want consumers to press their governments, food and water suppliers and above all their fellow consumers to insist upon.

We want to see all European states adopt a clearer approach to the use of pesticides and all agrochemicals. This should include:

i. a commitment to a progressive reduction of pesticide use.

ii. encouragement of those farming methods which are not dependent upon agrochemicals.

We want to see at least 20% of European agricultural production 'organic' or 'biological' by the year 2000. If Sweden can dramatically drop its pesticide use without unacceptable loss of productivity, then why cannot other European countries follow its lead?

We want to see a phasing out of those subsidies which are responsible for hazards to health, safety and the environment.

We want all European countries to adopt a tougher system of evaluation of the agrochemical industry and its products. The benefit of the doubt should always go to the consumer, environmental and public health interest. The principle should be: 'if in doubt, leave it out'. Any agrochemical subject to scientific review should be removed from use immediately.

We want all manufacturers and users of agrochemicals to take responsibility for their use throughout the lifespan of the product. This is known as the 'cradle to the grave' principle.

Clean, safe food, pure air and drinking water are basic human rights. They should be free from contamination by pesticides and other residues.

It is time for Europe's consumers to exert pressure on the market, suppliers of food and drink, and their governments to help get farmers and growers off the agrochemical treadmill.

Food

- there should be a programme of development of test procedures to enable a wider range of pesticides to be determined speedily and accurately, together with more broadly based multi-residue techniques.
- there should be continued monitoring of DDT in breastmilk and research to ascertain if intakes at high levels have effects adverse to the health of children. This should be undertaken within a context of full encouragement for mothers to breast-feed their children.
- governments should set up or increase existing public analytical services to allow members of the public improved access to information about pesticide levels in food. Levels of pesticides should be surveyed nationally for each type of food and the results published frequently so that people can tell which foods have high levels of residues.
- Maximum Residue Limits should be set for an increased range of pesticide and commodity combinations, to take account of growing concern about the 'chemical cocktail' effect. There should be an increased programme of research on the effect of processing on pesticide residues in food.
- agriculture ministries should produce bulletins listing pesticides currently in heavy use on particular products, for the benefit of local authority enforcement officers, and there should be liaison between Agriculture, Health and Environment officials, and local authority food enforcement officers.

Water

- the pattern of both agricultural and non-agricultural use of pesticides has to be established, together with a programme of monitoring of pesticides in water, taking into account seasonal variations of levels in water, and the local hydrogeological environment.
- the lack of cheap, readily available techniques for the analysis and monitoring of pesticides in water should be remedied as a priority; and it should be a condition of future registration of an active agrochemical ingredient that such a technique is available for its detection in water.
- research is urgently required into the environmental fate and

degradation characteristics of pesticides and metabolites in soil and water and oceans.

- powers should be taken by government for local control of pesticides usage to prevent contamination of water supplies.
- there should be identification and monitoring of landfill and other sites of potential health hazard from pesticides; criteria and recommendations for cleaning up waste sites should be developed, together with appropriate soil quality standards.

Health

- the availability of pesticides for domestic and household use should be reduced, to minimise hazards particularly to children and particularly from Paraquat.
- there should be more research into the chronic adverse health effects of pesticides and particularly in the area of cancer, reproductive hazards, neurotoxicity and immunotoxicity.
- body levels of pesticides and other toxic substances should be surveyed nationally by fat biopsy, blood sampling and other techniques. When public concern is expressed about pesticides, governments should be able to provide information about levels and exposure of specific regions and population groups.
- there is a need for the development of specific antidotes to non-insecticidal pesticides to treat acute poisonings. The provision of such antidotes should be a condition of registration.
- more training in toxicology should be available for doctors and all relevant medical professionals to assist in the recognition and management of pesticide poisoning.
- there should be evaluation of the merits of animal testing with a view to stimulating the development of other toxicological techniques.
- full toxicological data on pesticides should be publicly available, together with the criteria for evaluating health hazards.

Product liability

- consumers' rights should be given greater protection under national product liability laws. Where countries have no product liability laws or regulations, these should be introduced. In the European Community, current national implementation of European Community product liability laws should include agricultural and food products sometimes excluded.

- consumers should have the right to take actions against companies and governments that condone unsafe use of agrochemicals and which subsequently create unnecessary risks to human, animal and environmental health and welfare.
- damages in actions should be unlimited, and there should be no-fault compensation.
- agrochemical companies should be responsible for their products from cradle to grave, from the point of production to the effects and consequences of use.

Retailers and caterers

Retailers, caterers and all sellers of food and drink should:

- publish and make their procurement specifications available, and in particular reduce their insistence on blemish-free produce, concomitant with the need to produce safe food.
- review their pesticide policies and consider how they can reduce overall pesticides use and encourage the growing of pesticide-free produce.
- encourage the availability and sale of pesticide-free food and drink by reviewing their pricing policies, and where possible cross-subsidising 'organic' or 'biological' produce. They should do whatever is possible to lower the price difference between 'organic' or 'biological' and agrochemically produced produce.
- press for greater access to government and commercial information on the toxicity of pesticides and press for research into cheaper and more available methods of residue analysis.
- introduce labelling schemes for produce to indicate pesticide treatment pre- and post-harvest.

Farmers and growers

- farmers and growers should reduce their overall usage of pesticides.
- there should be an end to the practice of aerial spraying of pesticides on arable crops.

- there should be introduced provisions for the giving of mandatory notice of spraying, together with the provision at public information points of health and safety information for neighbours, occupiers, and other parties who may be affected by agricultural or amenity spraying, or timber treatment.
- farmers and growers should press for more information to be made available to them from manufacturers and government about the toxic effects and safe use of pesticides.
- resources should be made available for research into Integrated Pest Management, organic, and other systems of agriculture and pest control.
- governments should increase each year the proportion of support for non-agrochemical-reliant forms of food production.
- there should be a series of simple advice leaflets for gardeners and consumers on pesticides and all agrochemicals, what they are, their hazards and alternatives. These should be part of major public education and information programmes on agrochemicals and their use.

Occupational use

- agriculture or health and safety authorities should publish safety assessments and emergency plans, and air, water and sewage pollution figures, for all major pesticide manufacturing and formulation facilities.
- records should be published by manufacturers and formulators of the production quantities of all active ingredients and formulated products.
- there should be implemented a scheme for the collection and disposal of pesticides whose approvals are revoked or severely restricted, to reduce the public health risk of otherwise hazardous disposal.
- research should also concentrate on high risk and unstudied exposure groups, particularly workers in manufacture of pesticides, agriculture and application, and women.
- support and funding should be given to programmes to improve application technology and protective clothing, with a view to reducing overall use of pesticides and operator hazard.

International

- financial, technical, educational and regulatory support should be forthcoming from industrialised countries to assist in the regulation of pesticides; their safe use; the development of 'alternatives'; and the promotion of Integrated Pest Management programmes.
- the seven pesticides identified by the international Pesticides Action Network (PAN) and the Pesticides Trust in the UK – Carbofuran, Dichlorvos, Methamidophos, Methomyl, Methyl Parathion, Monocrotophos, Phosphamidon – should be included forthwith in 'Prior Informed Consent' procedures as these products have serious adverse health effects in the Third World. Under 'Prior Informed Consent' procedures, countries should be informed about potential and actual hazards of agrochemicals they are importing.
- the Pesticides Action Network (PAN) 'Dirty Dozen' pesticides – Campheclor, Chlordane/Heptachlor, Chlordimeform, DBCP, DDT, the 'drins' (Aldrin, Dieldrin, Endrin), EDB, HCH/Lindane, Paraquat, Ethyl Parathion, Pentachlorophenol/PCP and 2,4,5-T – should be withdrawn.
- both importing and exporting countries should recognise a shared responsibility for international trade and exporting countries should be willing to take necessary action – if necessary by export control – to prevent despatch of unwanted chemicals for which consent has not been given.
- the export of pesticides banned, severely restricted, withdrawn, or not registered for use, for reasons of health or environment in their country of origin, should be prevented.
- pesticides that are restricted in use, and listed in the World Health Organisation's (WHO) Class Ia ('extremely hazardous') or WHO Class Ib ('highly hazardous') pesticides, should be subject to Prior Informed Consent procedures before export takes place.
- the use of pesticides in projects supported by governments, bilateral and multilateral aid agencies should be countenanced only within the context of sustainable agriculture and the use of Integrated Pest Management. The appropriate guidelines for procurement and use should be monitored and enforced by all governments and agencies worldwide.
- the use of internationally acceptable standards of health and safety should be encouraged in the manufacture of hazardous chemicals by developing countries.

The European Community and 1992

We recognise that the map of Europe is changing and that in particular the removal of barriers to trade within the European Community (EC) poses both opportunities and threats for improvements to health. EC countries have a special responsibility to ensure that harmonisation of standards is to the highest and not the lowest common denominator. We call upon EC countries to undertake the following:

- member States should seek to harmonise registration of pesticides, ensuring the highest standards of public and environmental health.
- member States should develop a standardised means of analysing and reporting pesticide residues in food and water, and reporting the results.
- member States should press for a social, environmental and economic audit into the use of pesticides, with a view to reducing the overall use of pesticides.
- member States should develop a policy for the use of pesticides within the context of sustainable agriculture and Integrated Pest Management.
- member States should press for a levy on pesticides to fund research into alternatives and pollution control.
- member States should support the funding and development of the proposed European Environmental Protection Agency.
- member States should support measures to implement export control of pesticides not registered for use within the Member States, and to implement Prior Informed Consent procedures.

All European governments

- all governments should establish their own independent agency to be responsible for the national registration and monitoring of pesticides, with the resources to commission independent testing of active ingredients.
- no pesticide should be approved for use unless it can be shown to be safe according to current standards.
- sufficient resources should be devoted by government to reviewing the toxicity of older pesticides. The European Community should complete by 1995 all pesticides whose

clearances for use were granted before the implementation of the Plant Protection Directive and should include the publication of data gaps identified for all such pesticides.

- greater resources need to be devoted to Ministries of Agriculture, Health and Environment and local authorities to ensure proper monitoring and enforcement of safe pesticide use.
- a policy of reduction of use of pesticides should be introduced, with funding for research into alternative farming systems.
- consumer and environmental organisations should be represented in decisions taken about the use of pesticides. These organisations should be represented on relevant decision-making bodies.
- scientific committees which approve and monitor agrochemicals should contain at the very least a majority and ideally be wholly filled with people independent from the agrochemical industry.

National legislation

- a positive duty should be laid on manufacturers, formulators and importers of disclosure to government of any and all adverse effects of active ingredients; and to supply regular updated data packages for examination.
- 'polluter pays' legislation should be introduced to support enforcement actions by government agencies and ministries; and compulsory environmental insurance of manufacturers and formulators should be introduced, with the establishment of a 'Superfund' to assist in dealing with toxic waste.
- in the European Community, there should be speedy implementation of access to information promised under European Directives in order that workers, consumers and advisers can be aware of health and safety, environmental and other effects of pesticides, including data gaps.
- all pesticides should be classified as substances hazardous to health and brought within appropriate national health and safety legislation and regulations.
- records should be published of the import and export of pesticides that are banned, severely restricted or not approved by governments, and powers under European Directives should be invoked for this purpose.

Useful Addresses

Independent bodies

Association of Public Analysts: contact via your local Council.

Consumers' Association, 2 Marylebone Road, London NW1 ADX.

Friends of the Earth, 26–28 Underwood Street, London N1 7JQ.

Genetics Forum, 258 Pentonville Road, London N1 9JY.

Hazards, PO Box 199, Sheffield, S1 1FQ (for local group addresses).

Henry Doubleday Research Association, Ryton Gardens, Ryton, Coventry, Warwickshire.

Institution of Environmental Health Officers, Chadwick House, Rushworth Street, London SE1.

London Hazards Centre, Headland House, 308 Grays Inn Road, London WC1X 8DS.

Parents for Safe Food, Britannia House, 1–11 Glenthorne Road, London W6 0LF.

Pesticide Action Network (PAN) Europe, c/o Pesticides Trust.

Pesticide Exposure Group of Sufferers (PEGS), 10 Parker Street, Cambridge CB1 1JL

Pesticides Trust, 23 Beehive Place, London SW9 7QR.

Soil Association/Organic Growers' Association/British Organic Farmers, 86 Colston Street, Bristol BS1 5BB.

Women's Environmental Network, 287 City Road, London EC1.

TGWU, Transport House, Smith Square, London SW1.

BFAWLL Stanborough House, Great North Road, Stanborough, Welwyn Garden City, Herts.

GMB, Thorne House, Ruxley Ridge, Claygate, Esher, Surrey.

IPMS, 75–79 York Road, London SE1.

NUPE, 20 Grand Depot Road, Woolwich, London SE18.

Producers

British Agrochemical Association, 4 Lincoln Court, Lincoln Court, Peterborough, PE1 2RP.

British Crop Protection Council, 49 Downing Street, Farnham, Surrey.

Legislators/controllers

European Commission, 20 Rue de la Loi, Brussels, B-1040, Belgium.

Health and Safety Executive, Baynards House, Chepstow Place, London W2.

Ministry of Agriculture, Fisheries and Food, Nobel House, Smith Square, London SW1.

PIAP, c/o Health and Safety Executive.

Further Reading

There are many books and reports on pesticides. Here are some of the most useful and important.

General

Pesticides: The Hidden Peril. A Joint Trade Union Report on Pesticides Usage. Agricultural and Allied Workers' Trade Group, c/o TGWU, Transport House, Smith Square, London SW1P 3HB, 1986.

British Medical Association, *Pesticides, Chemicals and Health.* Edward Arnold, due summer 1991.

Bull, David, *A Growing Problem: Pesticides and the Third World Poor.* Oxfam, 1982.

Carson, Rachel, *Silent Spring.* Pelican Books, 1962.

Chetley, Andrew, *Cleared for Export: An Examination of the European Community's Pharmaceutical and Chemical Trade.* Coalition Against Dangerous Exports (CADE), 1985.

Cook, Judith and Chris Kaufman, *Portrait of a Poison: The 2,4,5-T Story.* Pluto, 1982.

Dover, Michael and Brian Croft, *Getting Tough: Public Policy and the Management of Pest Resistance.* World Resources Institute Study, WRI, 1735 New York Ave NW, Washington DC 20006, USA, 1984.

Dudley, Nigel, *This Poisoned Earth: The Truth about Pesticides.* Piatkus, 1987.

Dudley, Nigel, *Safety Never Assured: The Case Against Aerial Spraying,* Soil Association, 1985.

Dudley, Nigel, *Garden Chemicals.* Soil Association, 1986.

Food Magazine (quarterly), from: Food Commission (UK), 88 Old Street, London EC1V 9AR.

Gips, Terry, *Breaking the Pesticides Habit: Alternatives to 12 Hazardous Pesticides.* International Alliance for Sustainable Agriculture, Newman Centre, University of Minnesota, 1701 University Ave SE, Room 202, Minneapolis, MN 55414, USA, 1988.

Goldenman, Gretta and Sarojini Rengam, *Problem Pesticides, Pesticide Problems: A Citizens' Action Guide to the International Code of Conduct on the Distribution and Use of Pesticides.* IOCU/PAN, 1988.

Hazards (5 per year), from: PO Box 199, Sheffield S1 1FQ.

The Effects of Pesticides on Human Health ('The Body Report'). House of Commons Agriculture Committee (2nd Special Report), HMSO, 1987, 3 vols.

Hurst, Peter, Alistair Hay and Nigel Dudley, *Safe to Spray?* Pluto, due 1991.

Lees, Andy and McVeigh, Karen, *An Investigation of Pesticide Pollution in Drinking Water*. Friends of the Earth, 1988.

Toxic Treatments: Wood Preservative Hazards at Work and in the Home. London Hazards Centre, 3rd floor, Headland House, 308 Grays Inn Road, London WC1X 8DS, 1988.

MAFF, *Reports of the Working Party on Pesticide Residues*. (Food Surveillance Papers) HMSO,1982–5, 1985–8 and 1988–9.

Mott, Lawrie and Karen Snyder, *Pesticide Alert: A Guide to Pesticides in Fruits and Vegetables*. Natural Resources Defense Council/Sierra Club Books, 730 Polk Street, San Francisco, CA 94109, USA.

Monitoring and Reporting: The Implementation of the International Code of Conduct on the Use and Distribution of Pesticides. PAN International, 1987.

Pesticides Don't Know When to Stop Killing: The 'Dirty Dozen' Information Kit. Pesticides Education and Action Project (PEAP), PO Box 610, San Francisco, CA 94101, USA.

Pesticides News (Quarterly), from Pesticides Trust, see address under useful addresses.

Pesticides Trust, *Pesticides, Policy and People*, February 1991.

The FAO Code: Missing Ingredients. Pesticides Trust for PAN international, 1989.

Snell, Peter, *Pesticide Residues and Food: The Case for Real Control*. London Food Commission, 1986.

Watterson, Andy, *Pesticides and Food*. Greenprint, due 1991.

Watterson, Andy, *Pesticides Users' Health and Safety Handbook*. Gower Technical, 1988.

Weir, David and Mark Shapiro, *Circle of Poison: Pesticides and People in a Hungry World*. Institute for Food and Development Policy, 1885 Mission Street, San Francisco, CA 94103, USA, 1981.

Technical and reference sources

Hassall, K.A., *The Biochemistry and Uses of Pesticides: Structure, Metabolism, Mode of Action, and Uses in Crop Protection*. Macmillan, 2nd edn 1990.

MAFF, *Pesticides*. HMSO, annually.

The Agrochemicals Handbook. Royal Society of Chemistry, The University, Nottingham, 1987.

European Directory of Agrochemical Products. Royal Society of Chemistry, 1984, 4 vols.

UK Pesticide Guide, 1990, British Crop Protection Council.

The Crop Protection Directory. Edited and published by Elaine Warrell, 105 Lee Road, London SE3 9DZ, 1988.

Glossary

ACP	Advisory Committee on Pesticides
ADI	Acceptable Daily Intake
APA	Association of Public Analysts
ATB	Agricultural Training Board
BAA	British Agrochemical Association
BASIS	British Agrochemicals Standards Inspection Scheme
CDA	Controlled Droplet Application (see below)
CHEMAG	Chemicals in Agriculture Working Party of HSE
CHEMSAP	A sub-committee of CHEMAG, dealing with the pesticides approval process
CLA	Country Landowners' Association
COPA	Control of Pollution Act
COPR	Control of Pesticides Regulations
COSHH	Control of Substances Hazardous to Health
EC	European Community
EHO	Environmental Health Officer
EMAS	Employment Medical Advisory Service
EPA	Environmental Protection Agency (USA)
FAO	Food and Agriculture Organisation (UN)
FDA	Food and Drugs Administration (USA)
FEPA	Food and Environment Protection Act
GIFAP	Industrial Group of National Association of Pesticide Manufacturers
GMB	General, Municipal and Boilermakers' Union
HASAWA	Health and Safety at Work etc. Act 1974
HSE	Health and Safety Executive
IARC	International Agency for Research on Cancer
IPM	Integrated Pest Management
ISO	International Organisation for Standardisation
MAC	Maximum Admissible Concentration
MAFF	Ministry of Agriculture, Fisheries and Food
MRL	Maximum Residue Limit
NAAC	National Association of Agricultural Contractors
NIOSH	National Institute of Occupational Safety and Health
NOEL	No-Observed-Effect Level
NUPE	National Union of Public Employees
OEL	Occupational Exposure Level

PIAP	Pesticides Incidents Appraisal Panel
PSPS	Pesticides Safety Precautions Scheme (the voluntary body prior to COPR that approved pesticides)
SSC	Scientific Sub-Committee (recommends the ACP)
TGWU	Transport and General Workers' Union
WA	Water Act
WHO	World Health Organisation (UN)
WPPR	Working Party on Pesticide Residues (ACP)

Acaricide   Chemical that kills mites and ticks.

Active ingredient   The component(s) of a product that make it a pesticide.

Adjuvant   An inactive material that, when added to a pesticide, increases its efficiency.

Ames Test   A rapid laboratory test to detect mutagens.

Allergy   Hypersensitivity due to a chemical, aggravated by continued exposure.

Aphidicides   Chemicals that kill aphids (e.g. greenfly).

Biological control   Control of pests by natural enemies.

Carcinogenic   Causes cancer.

Chromosomes   The rows of genes.

Controlled droplet application   Makes spray drops about the same size.

Cholinesterase   An abbreviation of acetylcholinesterase, an enzyme which breaks down the chemical acetylcholine which transmits nerve impulses across the nerve gaps called synapses.

Common name   Well known name, generally adopted by the ISO.

Defoliant   Pesticide that causes leaves to fall.

Desiccant   Something which dries another thing out.

DNA   Deoxyribonucleic acid; The genetic building block.

Enzyme   A biological accelerator which speeds up body functions.

Entomologist   Scientist who studies insects.

Epidemiology   The study of patterns of disease among populations of people.

Foaming Agent   Stops spraydrift by forming a foam in the pesticide.

Fumigant   A chemical dispersed through the air in order to kill pests.

Fungicides   Pesticides used to kill fungi, e.g. in buildings or on farms.

Gene   Distinct part of the chromosome that determines features.

Genotoxic   Toxic to the gene.

Growth regulator   A chemical other than a nutrient which modifies plant growth.

Herbicides   Chemicals that kill unwanted plants.

Larvicide   Insecticide that kills larvae (grubs).

LD50   The dose required to kill 50% of the test animals in an experiment. It may be given by mouth, inhaled or injected, and is usually expressed as the weight of the chemical per unit weight of the animal.

Maximum Residue Limit The maximum concentration of a pesticide residue that is legally permitted or recognised as acceptable.

Molluscicide Substance that kills slugs and snails.

Mutagenic Causes changes in the genes which may or may not affect the next generation.

Narcosis Stupor or drowsiness, sometimes caused by pesticides.

Nematicide Substance that kills nematode worms.

Nematode Small unsegmented worm.

Neoplasm A growth of abnormal cells – a tumour.

Oncogenic Causes tumours.

Organochlorines Pesticides containing chlorine atoms. Sometimes called chlorinated hydrocarbons. DDT is an example.

Organophosphates Pesticides containing phosphorus, e.g. Malathion.

Ovicide Kills insect eggs before they hatch.

Percutaneous Through the skin.

Persistent Long-lasting and not rapidly broken down.

Phenoxyacetic acids Herbicides containing hydroxyl and acid substitutes, e.g. 2,4,5-T.

Potentiator Stimulates pesticides to work better.

Residual Remains active for some time.

Reproductive effects Include teratogenicity (see below), fertility, libido and menstruation.

Rodenticide Kills rodents such as mice and rats.

Sarcoma Cancer of connective tissue.

Synapse Nerve gap which nerve impulses have to pass.

Systemic effect Relating to or affecting the body as a whole, rather than just one part.

Systemic pesticide Enters plant and moves within it.

Tachycardia Increased heart rate. Can be caused by some pesticides.

Tank mix Several pesticides mixed in a spray tank prior to applying.

Teratogenic Causes birth defects.

Trade name Manufacturer's commercial name for a pesticide.

References

Place of publication for books is London unless otherwise stated.

Chapter 2

1. Lodeman, E. G., *The Spraying of Plants*, Macmillan, 1903.
2. *The Spraying of Plants*, op. cit.
3. Large, E. C., *Advance of the Fungi*, Cape, 1940.
4. *The Spraying of Plants*, op. cit.
5. Fletcher, W. W., *The Pest War*, Blackwell, Oxford, 1974.
6. Personal communication from M. Rossiter, director of *Fog of War* (ITV, 1990).
7. Green, M.B., *Pesticides: Boon or Bane?*, Environmental Studies, Elek, 1976.
8. Mooney, P., Personal communication.

Chapter 3

1. Ware, G., *Pesticides: Theory and Application*, W. H. Freeman, New York, 1983.
2. Latest DHSS figures available. Agriculture Select Committee, Vol. 2 p. 17, Table 2.
3. *Report of the Secretary's Commission on Pesticides and Their Relationship to Environmental Health*, US Government Printing Office, Washington DC, 1986.
4. Marquez Mayaudon, E. and A. Fujigaki Lechuga, C. A. Moguel and B. Aranda Reyes, 1968, *Salud Publica de Mexico*, Vol. 10, 3, p. 293, quoted in United Nations, op. cit.
5. Association of Public Analysts, Annual Report, 1985.
6. Working Party on Pesticide Residues (WPPR) report 1985-8, MAFF 1989, quoted in Taylor, J. and D., eds, *Safe Food Handbook*, Ebury, 1990, p. 138.
7. *Pesticides Incidents Investigated*, Annual Report, HM Agriculture and Factory Inspectorates, 1987–9, HSE.
8. House of Commons Agriculture Select Committee, *The Effects of Pesticides on Human Health*, Vol. 2, HMSO, 1987, p. 171, para 96.
9. Maclean, David, MP, Parliamentary Secretary, letter to Sir Christopher Booth, Chair of BMA working party on pesticides, 2 November 1990.
10. Earth Research Resources, *An Investigation into the Use of Pesticides*, Norwich City Council, September 1989.
11. Quoted on p. 249 in Davis, B. N. K. and C. T. Williams, 'Buffer Zone Widths for Honeybees from Ground and Aerial Spraying of Insecticides', *Environmental Pollution*, 63, 1990, pp. 247–59.
12. Figure on p. 253 in 'Buffer Zones Widths for Honeybees from Ground and Aerial Spraying of Insecticides', op. cit.
13. Lees, A. and K. McVeigh, *An Investigation of Pesticides Pollution in Drinking Water in England and Wales*, Friends of the Earth, November 1988, Table 1, p. 1.

14. *An Investigation of Pesticide Pollution,* op. cit.; see also report of Institute of Hydrology Study in FOE, *How Green is Britain?,* Hutchinson Radius, 1990, p. 78.
15. Varity, E., 'Weedkiller Seeps out of London Taps', *Today,* 28 June 1990.
16. *British Medical Journal,* 1970.
17. TGWU, GMB, NUPE, *Pesticides: The Hidden Peril.* A join trades union report on pesticide usage, London 1986.
18. Hazards, 23, 1989, pp. 8/9.
19. ENDS Bulletin, 1986, quoted in Pesticides Trust, *Pesticides, Policy and People,* February 1991.
20. Evidence to Agriculture Committee, *The Effects of Pesticides on Human Health,* Vol. 2, 1987, HMSO.
21. *The Effects of Pesticides on Human Health,* Vol. 2, p. 17, op. cit.
22. *Agrow,* no. 74, 24 October 1988, p. 6
23. 'UK chemical probe after 1000 deaths worldwide', *Safety,* September, 1982, p. 3.
24. Douglas, Jane, 'Poisons or Preservatives?', *Municipal Journal,* 15 September 1989, p. 26
25. 'Poisons or Preservatives?', op. cit., p. 26.
26. Racy, P. and S. Swift, 'The Residual Effects of Remedial Timber Treatments on Bats', *Biological Conservation,* 35, 1986, 205–14.
27. London Hazards Centre, *Toxic Treatments,* 1988, p. 8.
28. Anderson, L. et al, 'Parathion Poisoning from Flannelette Sheets', *Canadian Medical Association Journal,* 10 April 1965, Vol. 92, pp. 809–13.
29. Warren, M. C. et al, *Journal of the American Medical Association,* 1963, Vol. 1984, p. 286.
30. Hayes, W. J. *Toxicology of Pesticides,* Williams and Wilkins, 1975.
31. National Research Council, *Regulating Pesticides in Food: The Delaney Paradox,* National Academy Press, Washington DC, 1987, pp. 52-3, 107, 111, 114–6.
32. Scarborough, M. et al, 'Acute Health Effects of Community Exposure to Cotton Defoliants', *Archives of Environmental Health,* November/December 1989, Vol. 44, no. 6, pp. 355–9.
33. Ramalho, F. S. and F. M. M. Jesus, 'Evaluation of Electrodynamic and Conventional Insecticides against Cotton Boll Weevil and Pink Bollworm', *International Pest Control,* May/June 1989, pp. 56–60.
34. British Agrochemical Association, annual review and handbook 1988–9.
35. *Food Magazine,* Vol. 1, no. 2, summer 1988, p. 15.
36. 'An Investigation into the Use of Pesticides', op. cit.
37. 'Buffer Zone Widths for Honeybees from Ground and Aerial Spraying of Insecticides', *Environmental Pollution,* op. cit.
38. *The Environment Digest,* January 1990, no. 31.
39. WPPR 1985–8 report, quoted in Mabey, D., A. Gear and J. Gear, *Organic Consumers' Guide,* Thorsons, 1990, p 53.
40. Data from WPPR 1985–8 report, *Organic Consumers' Guide,* op. cit.
41. As note 40.
42. BBC Radio 4, *Farming Today,* 19 February 1989, transcript; also *Newcastle Evening Chronicle,* 20 February 1989.
43. *Farmers' Weekly,* 24 November 1989, p. 17.
44. *Agrow,* no. 112, 1 June 1990, p. 13.
45. United Nations, 1972, p. 148.

Chapter 4

1. Johnson, E., US EPA, personal communication to Pesticides Trust, 1990.
2. Data Requirements for Approval under the Control of Pesticides Regulations 1986 Appendix E Guidelines on the Classification and Labelling para 4.1.18.
3. Health and Safety Data for Garvox 3G (MAFF no. 15094), produced by Schering Agrochemicals Ltd, Hauxton, Cambridge, UK.
4. London Food Commission, *Food Adulteration and How to Beat It,* Unwin Hyman, 1988, p. 38.
5. British Crop Protection Council, Review 1989, p. 20.
6. Consumers' Association, *Which?,* October 1990.
7. Food Safety Act 1990, HMSO, Section 21.

Chapter 5

1. Ferrer, A. and J. Cabral, 'Epidemics Due to Pesticide Contamination of Food', *Food Additives and Contaminants,* 1989, Vol. 6, supplement no. 1, S95–S98; and Bull, D., *A Growing Problem,* Oxfam, 1982, p. 56.
2. Green, M. A. et al, 'An Outbreak of Watermelon-borne Pesticide Toxicity', *American Journal of Public Health,* 1987, Vol. 77, no II, pp. 1431–4.
3. Hirsch, G. et al, 'Report of Illnesses caused by Aldicarb-contaminated Cucumbers', *Food Additives and Contaminants,* 1987, Vol. 5, no. 2, pp. 155–6.
4. 'Epidemics Due to Pesticide Contamination of Food', *A Growing Problem,* both op. cit.
5. Major, C., *Food for Thought: A Response to Our Critics,* speech to a Friends of the Earth conference, *How Much Can We Stomach?,* ICI Agrochemicals, 1990, p. 6.
6. APA 1983, quoted in Dudley 1989.
7. Consumers' Association, *Which?,* October 1990.
8. MAFF Working Party on Pesticide Residues 1985–8, quoted in Consumers' Association, *Which?,* October 1990.
9. *Which?,* October 1990, op. cit.
10. Association of Public Analysts, *Survey of Pesticide Residues in Food,* APA, 1983.
11. The Daminozide story has been extensively covered, but is summarised in Table 2 of this book; also Parents for Safe Food, background briefing, May 1989; in American Consumers' Union, Consumer Reports, 1989; and in Motte et al, *Unacceptable Risk,* Natural Resources Defense Council, Washington DC, February 1989.
12. MAFF, Working Party on Pesticide Residues, report 1988–9, HMSO, 1990, and Health and Safety Executive, supplement to no. 8 1990 of the Pesticides Register, HMSO, September 1990.
13. Association of Public Analysts, Survey of Pesticides Residues in Food, op. cit.
14. MAFF, Working Party on Pesticide Residues, report 1982–5, HMSO, 1986.
15. *Which?,* October 1990, op. cit; Public Analyst's report to committee, Birmingham City Public Health and Environmental Protection Committee, 15 Feb 1991, report 3L.

16. London Food Commission, *Food Adulteration and How to Beat It,* Unwin Hyman, 1988, p. 98.
17. Op. cit., Table II, pp. 42–5.
18. British Medical Association, *Pesticides, Chemicals and Health,* report of the Board of Science and Education, October 1990, p. 98.
19. *Pesticides, Chemicals and Health,* op. cit.
20. MAFF, Report of Committee on Toxicity to Working PArty on Pesticide Residues, 1982–5, op. cit.
21. Report of Committee on Toxicity to Working Party on Pesticide Residues, 1982–5, op. cit.
22. Figure given for residue testing for UDMH fungicide residues to Parents for Safe Food and Friends of the Earth, MAFF press release, March 1990.
23. *Food Adulteration and How to Beat It,* op. cit., p. 102.
24. Stevens, R. A., private communication to T. Lang, Parents for Safe Food, from Southwark analytical and scientific services, 28 September 1990.
25. Citizens' Environmental Laboratory, National Toxics Campaign, 1168 Commonwealth Avenue, Boston, Massachusetts 02134, USA.
26. Snell, P., op. cit. pp. 98–100.
27. Food Safety Act, HMSO, 1990.
28. Quoted in Dudley, N., 'Pesticides Under Our Skin', draft manuscript, Soil Association and Parents for Safe Food, 1990.

Chapter 6

1 *Cancer Risk of Pesticides in Agricultural Workers,* American Medical Association Council on Scientific Affairs, JAMA, 1988, Vol. 260, no. 7, p. 959–66.
2. *Farmers' Weekly,* June 1990.
3. ENDS Report 1986.
4. Agriculture Committee, *The Effects of Pesticides on Human Health,* HMSO, 1987, Vol. 1, p. vii.
5. Op. cit., p. 294.
6. Body Report, op. cit., Vol. 2, p. 296.
7. British Toxicological Society: Memorandum to the Body Enquiry, p. 356.
8. Ibid, p. 356.
9. Weiss, 'Behaviour as an Early Indicator of Pesticide Toxicity', *Tox. and Ind. Health,* 1988, Vol. 4, no. 3, pp. 351–9.
10. Thomas, P. T. and R. V. House, 'Pesticide-Induced Modulation of the Immune System', in *Carcinogenicity and Pesticides,* ed. Ragsdale and Menzer, 1989, pp. 94–106.
11. Exon, J., N. Kerkvliet and P. Talcott, 'Immunotoxicity of Carcinogenic Pesticides and Related Chemicals', in *Jo. Env. Sci. Health,* 1987, C5 (1) 73–120; 'Pesticide-Induced Modulation of the Immune System', op. cit.
12. Fiore, M. C., H. C. Anderson et al, 1986 *Environ. Res.* 41, 633–645.
13. Sax, N. I., *Dangerous Properties of Industrial Materials,* Van Nostrand Reinhold; Snell, P., *Pesticide Residues – The Case for Real Control,* London Food Commission, 1986; Snell, P., 'Pesticides Can Damage Your Health – Official', in London Food Commission, *Food Adulteration and How to Beat It,* Unwin Hyman, 1988, Ch. 4; Watterson, A., *Pesticides: A User's Handbook,* Gower, 1988.

14. Fletcher, A., *Reproductive Hazards of Work,* Equal Opportunities Commission, Manchester, 1985.
15. *Science,* Vol. 186, p. 904.
16. *British Medical Journal,* 25 January 1975, p. 170.
17. IARC Monographs, IARC 1987.
18. Op. cit., p. 22.
19. Pesticides: International Food and Water Safety. G. Ekstrom, M. Akerblom. *Rev. Env. Cont. & Toxicol.* 1990, 114, 23–56.
20. *Regulating Pesticides in Food,* National Academy Press, Washington 1987.
21. See correspondence from Edward Groth to Bruce Ames, ACU, 7 December 1989.

Chapter 7

1. *Guardian,* 9 May 1990.
2. Revill, J., 'Danger Pesticide Found in Chocolate', *Mail on Sunday,* 21 January 1990.
3. *Food Magazine,* October/December 1989, p. 19.
4. *Food Magazine,* January/March 1990, p. 33.
5. Weeks, D. E., Bulletin of the World Health Organisation, 1967, Vol. 37, p. 499; quoted in Wolfe, H. R., 'Safety Problems Related to Transportation and Storage of Toxic Pesticides', in United Nations, *Developing Countries,* New York, 1972, pp. 145–51.
6. Gomez Ulloa, M., P. F. Velasco, H. Laverde de Fandino and M. E. Guerrero, 'Epidemiological Investigation of the Food Poisoning Which Occurred in the Municipality of Chiquinquira, Colombia, a Preliminary Report', Ministry of Public Health, Bogota, Colombia, 1967; quoted in *Developing Countries,* op. cit.
7. Hatcher, R., M. S. Dickerson and J. E. Peavey, United States Public Health Service, National Communicable Diseases Center, Morbidity and Mortality Weekly Report, 5 October 1968, p. 376; quoted in *Developing Countries,* op. cit.
8. *Developing Countries,* op. cit., p. 148.
9. *Evening Standard,* International Section, 20 June 1990.
10. e.g. Press Release 325/90, coinciding with Report of the WPPR 1988–9, MAFF and HSE.
11. Pesticides Trust, *The FAO Code: Missing Ingredients, Report on Prior Informed Consent,* prepared for Pesticides Action Network International, October 1989.
12. The notion of ecological dumping was outlined in the Declaration of Geneva, a petition from 30 non-governmental organisations to the General Agreement on Tariffs and Trade (GATT), 20 February 1990, European Ecumenical Organisation for Development, 174 Rue Joseph II, B-1040 Brussels; also available from Parents for Safe Food.
13. *The FAO Code: Missing Ingredients,* op. cit., p. 7.
14. *The FAO Code: Missing Ingredients,* op. cit., p. 41.
15. *The FAO Code: Missing Ingredients,* op. cit., p. 73.
16. *The FAO Code: Missing Ingredients,* op. cit., p. 25.
17. At the time of going to press, this process was still undecided. Agreement was there in principle; see Australian Consumers' Association, Pesticides Charter, Marrickville, Sydney, Australia, November 1990.

18. Hurst, P., A. Hay and N. Dudley, Pluto, forthcoming, 1991.
19. Global Pesticide Campaigner, vol 1, no 1, Oct 1990, PAN North America, San Francisco, USA.
20. *Lancet*, 17 December 1977, p. 1259; Intercontinental Press 1977, p. 1128 quotes Philips, who wrote it in a letter to OSHA, 12 September 1977 and *New York Times*, 27 September 1977. *New Scientist*, 1 September 1977. *OCAW Lifelines*, August 1977.
21. Yoko Nakamura of 2-245 Minamiyakata, Sakaemachi, Toyoake, Aichi, Japan, writing in *PAN*, Vol. 1, no. 1, Philippine Workers, Japanese Consumers and Banana Pesticides.
22. PAN Dirty Dozen Pesticides Fact Sheets.

Chapter 8

1. Gips, Terry, *Breaking the Pesticide Habit: Alternatives to 12 Hazardous Pesticides*, International Alliance for Sustainable Agriculture, pub. no. 1987-1, Appendix A.
2. Agrow 112, June 1990.
3. *Breaking the Pesticide Habit*, op. cit.
4. British Agrochemicals Association, Annual Review and Handbook, 1989/90.
5. Erlichman, J., *Guardian*, 10 December 1990.
6. Agrow, 112, June 1990, p. 1.
7. British Agrochemicals Association Annual Report, 1989/90.
8. *Breaking the Pesticide Habit*, op. cit.
9. 1987 figures given by Schering to Agrow, Pesticides Report, autumn 1990.
10. ADAS, *Arable Farm Crops*, 1988 Survey Report, 78, 1990, MAFF.
11. *Pesticides Register*, MAFF/HSE HMSO, no. 7, 1990, p. 2.
12. Letter to Tim Lang from the Minister's office, concerning the European Consumers' Pesticide Charter, July 1990; figures repeated in Food Safety Directorate Information newsletter, Ministry of Agriculture, Fisheries and Food, August 1990, p. 3.
13. *Farmers' Weekly*, 1 December 1989, Report of British Crop Protection Council.
14. 'UK Weeds – a Burning Issue?', Agrow, July 1990, no. 115, p. 10.
15. Agrow 107, 16 March 1990, p. 8.
16. Baron, Homer, Ciba-Geigy Report, 1990.
17. *Farmers' Weekly*, 9 December 1989.
18. Eurobarometer Survey, published in National Consumers' Council, *Consumers and the Common Agricultural Policy*, HMSO, 1988.

Chapter 9

1. Quoted in 'Exciting Utterances of the Year Awards', *Seattle Post Intelligencer*, 19 July 1990, p. A9.
2. Hardell, L and Axelson, 1986 *Phenoxyherbicides*; some comments on Swedish experiences in *Cancer Prevention* by C E Becker and M E Coye, Hemisphere Publishing, Washington D.C., 1986.
3. *Guardian Weekly*, 26 December 1982. Erlichman, J., 'Crop Spray Linked to Cancer', *Guardian*, 13 October 1990. Also in Gips, Terry, Breaking the Pesticide Habit:*12 Alternatives*, IASA, 1987., p. 126–8.

4. *Safe Use of Pesticides,* 20th Report of the WHO Expert Committee on Insecticides, Technical Report Series no. 513, Geneva, 1973, p. 18.
5. Epstein, Samuel, 'Corporate Crime', *Ecologist*, Vol. 19, no. 1, 1989.; Toxic Materials News 8 February, 1989, vol 16 no 6.
6. MAFF, Data Requirements under COPR 1986.
7. *Pesticide Manual,* British Crop Protection Council, any edition.
8. Agricultural and Allied Workers' Trade Group of the Transport and General Workers' Union, circular, 3 January 1985.
9. Consumers' Association, 'Pesticide Residues: A Consumer's Guide, *Which?* October 1990.
10 'Regulating Pesticides', *EPA Journal,* June 1984.
11. 'Crop Spray linked to Cancer', op. cit.
12. 'Exciting Utterances of the Year Awards', op. cit.
13. 'The Delaney Clause', Dateline Washington Column, *Consumers' Research*, January 1989, p. 4.
14. 'Regulating Pesticides', op. cit.
15. 'How Much Pesticides Do Consumers Eat?', *Consumers' Research*, February 1990, p. 4.
16. Consumers' Institute (N2), 'Pesticides in Cereals', *Consumer Food and Health*, April 1990, pp. 15–18.
17. Danish Technology Board, Consensus Document on Food Irradiation, final document, 22–24 May 1989, Copenhagen, 1989.

Chapter 10

1. Advisory Committee on Pesticides, Annual Report, 1988, HMSO, 1990.
2. Threshold Limit Value for year, Technical Data Note 2/year. HSE/HM Factory Inspectorate. Early versions had this statement, later versions (although often with the same levels) do not. Later versions called Occupational Exposure Limits Year. Guidance Note EH 40/Yr.
3. Documentation of Threshold Limit Values. American Conference of Government and Industrial Hygienists Inc., Cincinnati, Ohio, 5th edn, 1986.
4. Documentation of Threshold Limit Values, op. cit.
5. *A Comparison of PELs and TLVs to Health-based Exposure Limits Derived from the IRIS Database.* New Jersey Dept of Health, Occupational Health Service, 1988.
6. Information from Die Verbraucher Initiative at conference on 1992, Bonn, 20–22 April 1990.
7. Dr Kate Short of Sydney's Total Environment Centre and Peter Dingle, Dept of Environmental Science, Murdoch University, Perth, Western Australia are both researching this area.
8. See the excellent publications by the Total Environment Centre (ref. 6); also Conacher, J., *Pests, Predators and Pesticides,* Organic Growers' Association, 3rd edn, 1986, OGA, Western Australia.

Chapter 11

1. Pesticides Trust and PAN, *FAO Code: The Missing Ingredients,* PAN, 1989.
2. *European Community Controls on the Use of Pesticides on Foodstuffs.* Comments by Consumers in the European Community Group. CECG 90/11 Rev 1.

3. MAFF ACP 1990 membership list, 26 February 1991

4. Private communication to Tim Lang by member of ACP, May 1989.

5. *Pesticide Incidents Investigated in 1989/90*, report of HM Agricultural and Factory Inspectorates, HSE, 1990.

6. Ritchie, Mark, *The Ecologist*, GATT special issue, Vol. 20, no. 6, November/December 1990.

7. Ritchie, Mark, 'GATT, Agriculture and the Environment', *The Ecologist*, Vol. 20, no. 6, November/December 1990. Moore, Monica, 'GATT, Pesticides and Democracy', PAN, Vol 1, no. 1.

8. Barclay, B. Table in *Global Pesticides Campaigner, Pesticides Action Network North America*, Vol. 1, no. 1, October 1990, p. 14.

9. Finney, J. 'Where do we stand? Where do we go?', paper at 7th International Conference of Pesticide Chemistry, Hamburg, 5–10 August 1990.

Chapter 12

1. The latest is *Pesticides 1990: Approved under the Control of Pesticide Regulations 1986*, HMSO, 1990.

2. *Pesticides 1990 Pesticides Approved under the Control of Pesticides Regulations 1986*. Ref. Book 55, MAFF/HSE, HMSO, 1990.

3. *Pesticide Register*, is produced jointly by MAFF and HSE, and published by HMSO.

4. Green Coalition press release, 28 August 1990, Pesticides Trust and others.

5. HSE Prestel Information Service, December 1990, p. 575489a.

6. *Pesticide Register*, no. 9, 1990.

7. *Pesticide Register*, no. 9, 1990.

8. Cook, J. and C. Kaufman, *Portrait of a Poison: The 2,4,5-T Story*, Pluto, 1982.

9. Joint statement by Green Alliance, British Agrochemicals Association, TGWU, National Federation of Women's Institutes, Pesticides Trust and Friends of the Earth, 1989.

10. Advisory Committee on Pesticides Annual Report 1988, HMSO, 1990.

11. *Pesticide Register*. Monthly listing of approvals and government announcements, MAFF/HSE, HMSO.

12. *Food Magazine*, no 11, October– December 1990, p. 21.

13. HSE Technical Divison, Specialist Inspector Reports, no. 24, *Chemicals Handling at Fish Farms: A Survey*, p. 2.

Chapter 13

1. *Pesticides Incidents Investigated in 1989/90*, Report by HM Agricultural and Factory Inspectorate, July 1990, HSE.

2. *Independent*, 12 December 1990.

3. *Health and Safety: An Alternative Report*, IPMS, 1990.

4. MAFF/HSE, Pesticides Register, 10, 1990, p. 2.

5. MAFF/HSE, Pesticides Register, 10, 1990, p. 2.

6. *Lancashire Evening Post*, November 1990.

Chapter 14

1. Latest at time of going to press is *Pesticides Incidents Investigated in 1989/90,* report by HM Agricultural and Factory Inspectorates, HSE, 1990.
2. Barnes, J. M., 'Pesticide residues as hazards', *PAN,* Vol. 15, no. 1, March 1969, p. 2.
3. Naish, D., speech to UKASTA conference *Food in the 90s – the Role of Pesticides?,* Northampton, 8 March 1990.
4. MAFF food labelling survey in England and Wales, HMSO 1990.
5. Diehl, E., 'Does the Use of Fertilisers and Pesticides in Plant Production Impair the Wholesomeness of Foodstuffs of Animal Origin?', *Acta Alimentaria,* Vol. 15 (4), p. 337 (1986).
6. National Research Council, *Regulating Pesticides in Food,* National Academy Press, Washington, DC, 1987.
7. Robinson, C., *Hungry Farmers: World Food weeds and Europe's Response,* Christian Aid, London p. 5.
8. Robinson, C., op. cit., p. 13.
9. Robinson, C., op. cit., p. 17.
10. Moore Lappe, F. and J. Collins, *Food First: Beyond the Myth of Scarcity,* Ballantine, New York, 1978; George, S., *How the Other Half Dies,* Penguin, 1976; Wisner, B., *Power and Need in Africa,* Earthscan, 1988.
11. See Cannon, G. and A. Henderson, *Superbug,* Ebury, 1990, forthcoming.
12. Van den Bosch, R., *The Pesticide Conspiracy,* Prism Press, Dorchester, UK, 1980; quoted in Bull, D., *Pesticides – a Growing Problem,* Oxfam, 1982, p. 16.
13. *Pesticides – a Growing Problem,* op. cit., p. 4.
14. Correspondence between Professor Bruce Ames and Dr Edward Groth for the American Consumers' Union, autumn and winter 1989.
15. 'How much can we stomach?', conference on the use of chemicals in farming and fresh food production, report on Friends of the Earth conference, Midhurst and Haslemere, 30 September 1989, *Farmers' Weekly,* 6 October 1989, p. 22.
16. Letter from E. Groth on behalf of Consumers' Union to Professor Bruce Ames, 7 December 1989, pp. 1–25.
17. Report of the Working Party on Pesticide Residues 1988–9, Ministry of Agriculture, Fisheries and Food, and Health and Safety Executive, supplement to Issue 8 of the *Pesticides Register,* HMSO, October 1990.
18. As note 13.
19. Food Magazine, quoted in, Lang, T. et al, *This Food Business,* Channel 4 Education and *New Statesman* and Society supplement, July 1989, p. 5.
20. British Agrochemical Association, Annual Review and Handbook, 1989–90, p. 14.
21. British Medical Association, 'Pesticides, Chemicals and Health', report of a working party of the Board of Education and Science, October 1990.
22. Pesticides Trust, *The Control of Pests and Pesticides,* broadsheet, January 1990; also Lohse, J., S. Winteler and D. Susat, *Lindane and other Pesticides in the North Sea: A Reason for Concern,* Greenpeace International North Sea Research Report, 1990.
23. *The Control of Pests and Pesticides,* op. cit.
24. Sanders, T., in a BBCTV *London Plus* interview, 4 March 1986, quoted in Cannon, G., *The Politics of Food,* Century, 1987, p. 92.

25. 'How much can we stomach?', op. cit.
26. *The Pesticide Conspiracy,* op. cit. p. 136.
27. Kronfeld, Professor D., speech to European Parliament symposium, 1988.
28. For example, Coopers and Lybrand Deloitte, Organic Foods Report, Birmingham, 1990.
29. Major, C., speech to UKASTA, 8 March 1990, p. 2.
30. Agrow, no.110, 2 May 1990, p.11.
31. British Agrochemical Association, Annual Review and Handbook 1989/90, p. 15.
32. Parents for Safe Food, Pesticides Trust, Women's Environmental Network, Soil Association and others, European Consumers' Pesticide Charter, European Ecological Consumers Coordination, May 1990.
33. Verlander, N., Briefing on environmental labelling, Friends of the Earth, 1990, and FOE, *How Green is Britain?* Hutchinson Radius, 1990, Chapter 10.
34. Friends of the Earth, *The Limits of Green Consumerism,* 1989.

Chapter 15

1. Corbett, J., chairman's foreword, Annual Review and Handbook 1989–90, British Agrochemical Association, p. 3.
2. Quoted in Lappe, F. M. and J. Collins, *Food First,* Souvenir Press, 1980, p. 53.
3. Clutterbuck, C. and T. Lang, *More Than We Can Chew,* Pluto 1982.
4. Bridger, G. and M. Soisons, *Famine in Retreat,* Dent, 1970, p. 39.
5. NEDO, Direction for change in land use in the 1990s, Agriculture GDC, 1987.
6. Porritt, J., *Where On Earth are we Going?,* BBC Books 1990, Chapter 3.
7. Gardiner, D., *Financial Times,* 24 January 1991.
8. *Independent,* 24 August.
9. George, S. *How the Other Half Dies,* Penguin, 1976.
10. *Getting a Handle on Pesticides.* Trans-national Information Exchange, no. 15.

Chapter 16

1. Farquhar, I., 'The Organic Paradox', *Food Matters,* no. 5, March 1990, p. 7.
2. See Dudley, N., *This Poisoned Earth,* Piatkus, 1988, for some alternatives; also contact the Henry Doubleday Research Association, Ryton, Coventry, Warwickshire; *Gardening* from *Which?* has a gardening without chemicals book, 1990.
3. Nicol, Hugh, *Biological Control of Insects,* Penguin, 1943.
4. Van den Bosch, R. and P. Messenger, *Biological Control,* Intertext Books, 1973.
5. Ayres, P. and N. Paul, 'Weeding with Fungi', *New Scientist,* 1 September 1990.
6. MAFF, Food Labelling Survey, HMSO, 1990.
7. Gips, Terry, *Breaking the Habit. Alternatives to 12 Hazardous Pesticides,* International Alliance for Sustainable Agriculture, no. 1987–1.
8. Van den Bosch, R. and M. L. Flint, *Source Book on Integrated Pest Management,* R. Van den Bosch, *Pesticide Conspiracy,* Prism Press, 1978.
9. *Fighting Pests the Natural Way,* PAN Europe, 1988.
10. Community Hygiene Concern, 32 Crane Avenue, Isleworth, Middlesex.
11. Consumers' Association, *Which?,* October 1990; also Mabey, D. Gear, A. and J. Gear, *Organic Consumers' Guide,* Thorsons, 1990.

12. French Aird, Jackie, *Organic Control of Household Pests*, Melbourne, 1988, chapter on Plants for Pest Control; also Forsythe, T., *Organic Pest Control*, Thorsons, 1990.
13. Cawcutt, L. and C. Watson, *Pesticides: The New Plague*, Friends of the Earth, Fitzroy, Victoria, Australia.
14. Comments in a seminar on Sweden by representatives from Sweden's Farmers' Organisation and independent researchers, CLM Conference on GATT – *Agriculture and the Environment*, Amsterdam, 15 September 1990.
15. 'Progress in Swedish Pesticide Reduction', Agrow, no. 115, 13 July 1990, p. 8.
16. For more on pesticide reduction, see Hurst, P. et al, forthcoming, Pluto, 1991.
17. Rogers, Paul, *Safer Pest Control*, Kangaroo Press, Australia, 1986; see also Forsythe, T. (reference above).
18. *Toxic Treatments*, London Hazards Centre, 1988.
19. Ormerod, Eleanor, *Manual of Injurious Insects*, London, 1890, and *Handbook of Insects Injurious to Orchard and Bush Fruits*, London, 1989.
20. As note 19.
21. Forsythe, T., *Successful Organic Pest Control*, Thorsons, 1990.
22. Branwell, N., *How Green is Your Garden*, Thorsons, 1990.

Chapter 17

1. *Bioprocessing Technology*, June 1988, p. 2.
2. Hobbelink, H., 'New Hope or False Promise?', *Biotechnology and Third World Agriculture*, ICDA, 1987.
3. *Chemical Week*, 4 May 1988, p. 35.
4. Royal Commission on Environmental Pollution, 13 Report, 'The Release of Genetically Engineered Organisms to the Environment', HMSO, July 1989.
5. Erlichman, J., *Guardian*, 10 December 1990; also Genetics Forum, (see useful addresses).
6. Sun, M., *Science*, 1988, 242, p. 504.
7. *New Scientist*, 20 November 1986 and 14 April 1988.
8. Renwick, J. H. et al, *Teratology*, 1984, 30, 371.
9. Seligman, P. J. et al, *Arch. Dermatol*, 1987, 123, 1478.
10. 'New Hope or False Promise', op. cit. p. 13.
11. *New Scientist*, 11 August 1988.
12. As note 9.
13. Genetic Resources Action International, Seedling, May 1990.
14. King, D., E. Brunner, C. Wilbert, and T. Lang, Genetic Engineering, Briefing Paper, Parents for Safe Food and Genetics Forum, November 1990.
15. Finney, J. R., (R and D director, ICI Agrochemicals), 'World Crop Protection Prospects: Where Do We Stand? Where Do We Go?' 7th International Conference of Pesticide Chemistry, Hamburg, Germany, 5–10 August 1990.
16. Hazards 12, 1987, p. 9.
17. Hazards 31, 1990, p. 6/7.
18. National Consumer Council, *Consumers and the Common Agricultural Policy*, HMSO, 1988, figures updated.

References   251

Table 2: The Main Pesticides Found

1. Woo, Lai et al, *Chemical Induction of Cancer*, Vol. IIIc, Academic Press, San Diego, California, 1988.
2. ACP Annual Report 1989, HMSO, 1990.
3. Olson et al, *Arch. Environ. Contam. Toxicol.*, 1987, 16, 433–9.
4. Fiore et al, *Environ. Res.*, 1986, 41, 633–45.
5. ACP Annual Report, 1987, HMSO, 1989.
6. WHO Environmental Health Criteria 64, 1986.
7. EPA Fact Sheet 147, October 1987.
8. As note 11.
9. Patty, *Industrial Hygiene and Toxicology*, 3rd edition, John Wiley, Chichester, 1982.
10. ACP Annual Report 1984, HMSO, 1986.
11. Health Advisory, EPA, USA, September 1987.
12. *Scientific Reviews of Soviet Literature on Toxicity and Hazards of Chemicals*, no. 18, IRPTC, Geneva, 1982.
13. ENDS Report no. 183, 1990.
14. *IRPTC Bulletin*, Vol. 10/1, March 1990.
15. ACP Annual Report 1988, HMSO, 1989.
16. As note 19.
17. ACP Annual Report 1989, HMSO, 1990.
18. As note 21.
19. *WHO Environmental Health Criteria* 64, 1986.
20. EPA Fact Sheet 21, EPA, 1985.
21. Grover et al, 'Carbaryl – a Selective Genotoxicant', *Env. Polln.*, 1989, 58, 313–23.
22. As note 24.
23. Farage-Elewar, M., 'Enzyme and Behavioural Change in Young Chicks as a Result of Carbaryl Treatment', *Jo. Toxicol. and Env. H.*, 1989, 26, 119–31.
24. Garnett, Stack, et al, *Mutat. Res.*, 1986, 168, 301–25.
25. Noo, Lai et al, *Chemical Induction of Cancer*, Vol. IIIC, Academic Press, San Diego, California, 1988.
26. *Pesticide and Toxic Chemical News*, 1985, 13, (48), 14.
27. EPA Fact Sheet 1986.
28. *WHO Environmental Health Criteria* 64, 1986.
29. EPA Fact Sheet 197, January 1989.
30. As note 33.
31. *IARC Monographs on Estimation of Carcinogenic Risk*, 1987.
32. As note 35.
33. WHO Environmental Health Criteria 34, 1984.
34. As note 37.
35. *Farmers' Weekly*, 28 September 1990.
36. EPA Fact Sheet 36, 1986.
37. *WHO Environmental Health Criteria* 82, 1989.
38. EPA Fact Sheet 94, 2 September 1988.
39. Johnson, E. S., 'Association Between Soft Tissue Sarcomas, Malignant Lymphomas and Phenoxy Herbicides/Chlorophenols: Evidence from Occupational Studies', *Fund. and App. Toxicol.*, 1990, 14, 219–34.

40. Sharp, D. and B. Eskenazi, 'Delayed Health Hazards of Pesticide Exposure', *Ann. Rev. Pub. Health,* 1986, 7, 441–71.
41. As note 38.
42. EPA Press Release, 1 February 1989.
43. *Position Document on the Risk to Consumers from Daminozide,* MAFF, November 1989.
44. *Scientific Reviews of Soviet Literature on Toxicity and Hazards of Chemicals* 39, IRPTC, 1983.
45. *WHO Environmental Health Criteria* 83, 1989.
46. EPA Fact Scheet 96, 1 December 1988.
47. As note 46.
48. Robertson, J. and C. Mazzella, 'Acute Toxicity of the Pesticide Diazinon to the Freshwater Snail *Gillia altilis*', *Bull. Environ. Contam. Toxicol.,* 1989, 42, 320–4.
49. *WHO Environmental Health Criteria* 79, 1989.
50. *Toxicology and Carcinogenesis Studies of Dichlorvos,* US NTP, 1989.
51. *Pesticides News,* 6 Novermber 1989.
52. US National Toxicology Programme, 1983.
53. EPA Fact Sheet 16, 1983.
54. As note 53.
55. WHO Environmental Health Criteria 90, 1989.
56. ACP Annual Report 1989, HMSO, 1990.
57. *Pesticide Residues in Food,* FAO/WHO, 1989.
58. EPA Fact Sheet 167, May 1988.
59. As note 58.
60. EPA Fact Sheet 173, June 1986.
61. ACP Annual Report 1988, HMSO, 1989.
62. ACP Annual Report 1989, HMSO, 1990.
63. As note 62.
64. Magos, L., 'Thoughts on Life with Untested and Adequately Tested Chemicals', *Brit. Jo. Ind. Med.,* 1988, 45, 721–6.
65. EPA Fact Sheet 73, September 1985.
66. *Pesticides in the North Sea,* Greenpeace International, 1990.
67. EPA Fact Sheet 152, January 1988.
68. As note 67.
69. EPA, Notice of Preliminary Determination to Cancel Certain Registrations, Federal Register, 20 December 1989.
70. ACP, *Position Document on Consumer Risks Arising from the Use of EBDCs,* January 1990.
71. As note 69.
72. As note 69.
73. As note 70.
74. WHO Environmental Health Criteria 78, 1988.
75. As note 69.
76. EPA Fact Sheet 208, 1989.
77. As note 76.
78. EPA Fact Sheet 192, December 1988.
79. EPA Fact Sheet 98, 1986.
80. Dudley N., *Methyl Bromide and the Ozone Layer,* Earth Resources Research, 1990.

81. EPA Fact Sheet 131, 1986.
82. US National Toxicology Program, quoted in Ragsdale and Menzer, eds, *Carcinogenicity and Pesticides*, 1989.
83. *Recognition and Management of Pesticide Poisoning*, EPA, 1989.
84. WHO Environmental Health Criteria 71, 1987.
85. As note 84.
86. As note 83.
87. WHO Environmental Health Criteria 86, 1989.
88. EPA Fact Sheet 23, 1984.
89. ENDS Report 1990, 183.
90. Letter, 26 September 1989, ACP Approval Holders.
91. WHO Environmental Health Criteria 42, 1984.
92. ACP Annual Report, 1988.
93. ACP Annual Report, 1989, HMSO, 1990.
94. Letter, DoE to Chief Executives of Water Authorities, 1 March 1989.
95. WHO Environmental Health Criteria 63, 1986.
96. As note 95.
97. As note 95.
98. EPA, Notice of Preliminary Determination to Cancel Certain Registrations, Federal Register, 20 December 1989.
99. ACP Position Document on Consumer Risks Arising From the Use of EBDCs, January 1990.
100. IRPTC, *Reviews of Soviet Literature on Toxicity and Hazards of Chemicals*, no. 31, UNEP, 1983.
101. As note 98.

Table 4: The Symptoms of Pesticide Poisoning

1. Sources for symptoms described are:
   Watterson, A., *Pesticide Users' Health and Safety Handbook*, Gower Technical, 1988.
   Timbrell, J.A., *Introduction to Toxicology*, Taylor & Francis, 1989.
   Division of Occupational Health & Pollution Control, *Agricultural Chemicals: A Synopsis of Toxicity*, New South Wales, Australia, n.d.
   Hayes, W.J., *Toxicology of Pesticides*, Williams & Williams, 1975.
   Kay, K., 'Toxicology of Pesticides', *Environmental Research*, Vol. 6, 202–43, 1973.
   Natoff, I., 'Organophosphorous Pesticides: Pharmacology', *Progress in Medicinal Chemistry*, Vol. 8, 1–39, Butterworth, London, 1971.
2. British Medical Association, *Pesticides, Chemicals and Health*, Report of the Board of Science and Education, October 1990, BMA, to be published as a book by Edward Arnold in July 1991.
3. London Hazards Centre, *Toxic Treatments: London Hazards Centre Handbook*, 1988.

Index

See also the alphabetical list of food between pages 157–181 and pesticides between pages 182–213